The Ghost in the Telescope

The Ghost in the Telescope is an insider's account of the Herschel Space Observatory, which was launched to answer two of the biggest questions in astronomy: How were the stars and the galaxies born?

Written in an engaging manner for a general audience, this book tells the stories of the telescope itself, the discoveries it made, and the engineers and astronomers who built and used it.

This book, based on the author's own experience and interviews with the key astronomers and engineers, tells the story of the mission, from its original concept on a piece of paper in Venice to the moment after the end of the mission when the engineers had to decide whether to crash the spacecraft into the Moon. Containing some of the most spectacular pictures ever taken of the universe, this book describes all the major discoveries made with the telescope. It also gives an account, accessible to anyone without previous scientific knowledge, of the latest research into the births of stars and galaxies.

This book may interest anyone who is curious about astronomy, space missions and how astronomy is done in practice. It is designed to be easy to read and does not require any previous scientific background.

Stephen Eales spends his time trying to understand the births and life cycles of galaxies, using telescopes all over the world and in space. He is the author of over 300 peer-reviewed papers and, in 2016 he was awarded the Herschel Medal by the Royal Astronomical Society for his contributions to observational astrophysics.

The Ghost in the Telescope
The Story of the Herschel Space Observatory

Stephen Eales

CRC Press
Taylor & Francis Group
Boca Raton London New York

CRC Press is an imprint of the
Taylor & Francis Group, an **informa** business

Front cover image: NASA

First edition published 2026
by CRC Press
2385 NW Executive Center Drive, Suite 320, Boca Raton FL 33431

and by CRC Press
4 Park Square, Milton Park, Abingdon, Oxon, OX14 4RN
CRC Press is an imprint of Taylor & Francis Group, LLC

© 2026 Stephen Eales

Reasonable efforts have been made to publish reliable data and information, but the author and publisher cannot assume responsibility for the validity of all materials or the consequences of their use. The authors and publishers have attempted to trace the copyright holders of all material reproduced in this publication and apologize to copyright holders if permission to publish in this form has not been obtained. If any copyright material has not been acknowledged please write and let us know so we may rectify in any future reprint.

Except as permitted under U.S. Copyright Law, no part of this book may be reprinted, reproduced, transmitted, or utilized in any form by any electronic, mechanical, or other means, now known or hereafter invented, including photocopying, microfilming, and recording, or in any information storage or retrieval system, without written permission from the publishers.

For permission to photocopy or use material electronically from this work, access www.copyright.com or contact the Copyright Clearance Center, Inc. (CCC), 222 Rosewood Drive, Danvers, MA 01923, 978-750-8400. For works that are not available on CCC please contact mpkbookspermissions@tandf.co.uk

Trademark notice: Product or corporate names may be trademarks or registered trademarks and are used only for identification and explanation without intent to infringe.

For Product Safety Concerns and Information please contact our EU representative: GPSR@taylorandfrancis.com
Taylor & Francis Verlag GmbH, Kaufingerstraße 24, 80331 München, Germany.

Library of Congress Cataloging-in-Publication Data
Names: Eales, Stephen, author.
Title: The ghost in the telescope : The story of the Herschel Space
Observatory / Stephen Eales.
Description: First edition. | Boca Raton, FL : CRC Press, 2025. | Includes
bibliographical references and index.
Identifiers: LCCN 2025005439 (print) | LCCN 2025005440 (ebook) |
ISBN 9781032049496 (hardback) | ISBN 9781032043821 (paperback) |
ISBN 9781003195290 (ebook)
Subjects: LCSH: Herschel Space Observatory (Spacecraft) | Orbiting
astronomical observatories. | Space telescopes. | Stars–Formation. |
Galaxies—Formation.
Classification: LCC QB500.267 .E35 2025 (print) | LCC QB500.267 (ebook) |
DDC 522/.2919—dc23/eng/20250319
LC record available at https://lccn.loc.gov/2025005439
LC ebook record available at https://lccn.loc.gov/2025005440

ISBN: 978-1-032-04949-6 (hbk)
ISBN: 978-1-032-04382-1 (pbk)
ISBN: 978-1-003-19529-0 (ebk)

DOI: 10.1201/9781003195290

Typeset in Minion Pro
by codeMantra

For Keirsten - gratitude and all my love

Contents

Note to the Reader

ONE OF THE ASTRONOMERS I interviewed before writing this book said that he had been part of seven space missions and *Herschel* had been the most fun. One of the reasons I wrote it is that I didn't want the fun, dedication and general human craziness of the hundreds of scientists and engineers who were such an important part of the mission to be lost. This book is not an oral history, but it is partly based on interviews with some of the key *Herschel* scientists and is made up of all our personal stories. They are all true. One of my other goals was to describe some of the discoveries made with the telescope, including not just the big ones but also some of the smaller ones that may have moved the dial slightly in some research field that probably only a few hundred people around the world care about. So if you are not prepared to be excited about interstellar dust, this book was probably a bad choice. I also hope that this book gives a sense of what it felt like for an astronomer to be caught up in a space mission in the early twenty-first century.

Guide to Acronyms and Astronomers' Jargon

ALMA: The Atacama Large Millimetre Array in the Atacama Desert in Chile. By combining the signals from 66 separate dishes, it can take pictures with even more detail than *Hubble* and the *James Webb Space Telescope*. It's the toughest of all telescopes for getting observing time – only about 10% of proposals are successful.

CO: Carbon monoxide – after molecular hydrogen the second (or third) most common molecule in interstellar space (see Chapter 14).

Celestial equator: See item on the celestial sphere

Celestial poles: See item on the celestial sphere

Celestial sphere: The night sky looks like the inside of a sphere – the celestial sphere. The celestial poles are where it is intersected by a line passing through the Earth's poles, and the celestial equator is the projection of the Earth's equator. As astronomers, we measure positions on the celestial sphere using coordinates, right ascension and declination, which are very similar to latitude and longitude.

Core: A particularly dense part of a giant molecular cloud that will eventually collapse to form a star or stars.

Dark energy: One of the two big mysteries of modern astronomy. We know the expansion of the universe is accelerating, so there must be something causing the acceleration. Astronomers have called this something 'dark energy', but the cute name doesn't mean we have any real understanding of what it is.

Dark matter: The other big mystery of modern astronomy. Even though we can't see it, we know it is there because of its gravitational effect. We don't know what it is made of, although we know it is not made of regular stuff, the protons and neutrons that make up the stars, planets and us.

Ecliptic plane: The orbital plane of the planets. Its intersection with the celestial sphere (see item) is at an angle to the celestial equator because the Earth's axis is not perpendicular to the orbital plane of the planets.

Electromagnetic waves (electromagnetic spectrum): Waves that travel through the electromagnetic field. Radio waves, X-rays, infrared radiation, gamma rays, ultraviolet radiation, submillimetre radiation, visible light – all are just electromagnetic waves with different ranges of wavelength. Every astronomer has their personal definitions of these wavelength ranges, although mine are the correct ones.

ESA: The European Space Agency, the European equivalent of NASA.

ESOC: The European Space Operations Centre in Darmstadt Germany. The ESA equivalent of NASA Mission Control in Houston.

ESTEC: The European Space Research and Technology Centre in Noordwijk, the Netherlands – the design centre for all ESA space science missions.

Far-infrared waveband: Electromagnetic waves with wavelengths between 10 micrometres and 100 micrometres. Interstellar dust emits radiation in both the submillimetre and far-infrared wavebands, with warmer dust emitting more radiation in the latter waveband and vice versa.

Flux: The strength of an object's radiation as observed on Earth. Not to be confused with the object's luminosity, which is essentially its power wattage – how much radiation it is emitting rather than how strong the radiation appears on Earth.

Galactic poles: The points where the axis of the Galaxy's disk meets the celestial sphere (see item).

GT: Guaranteed Time, the observing time on a telescope that is granted to the team that has built a camera or one of its other instruments for a telescope.

HIFI: One of the three instruments on *Herschel*. Unlike the other two, HIFI didn't take pictures but spectra (see item). The letters stand for Heterodyne Instrument for the Far-Infrared.

ISO: *Infrared Space Observatory*, another ESA space telescope, which made observations in the infrared waveband between 1995 and 1998.

JWST: The *James Webb Space Telescope*. Launched in December 2021, the JWST is the successor to *Hubble* with a mirror roughly nine times larger in area.

Lagrangian Points: The five in most instruments today special points at which a small object can orbit in a constant pattern with two larger masses. In the Earth-Sun system, the second Lagrangian point is in direct line with the Earth and the Sun, on the opposite side of the Earth from the Sun. An object at L2 orbits the Sun in the same time it takes the Earth to orbit the Sun, so the three objects stay in a straight line.

LMC: Large Magellanic Cloud, one of the two closest galaxies to our own, with the other being the Small Magellanic Cloud.

Luminosity: Essentially an object's power wattage – how much radiation it is emitting. Not to be confused with its flux, the strength of the radiation as measured on Earth, which depends on the object's distance.

Noise: Irregular fluctuations that impede the detection of a signal. In an astronomical instrument, the signal is the radiation from a star or galaxy, which in most instruments today generates an electrical current or voltage in the detector. The noise is the irregular fluctuations in the current or voltage that have nothing to do with the star or the galaxy. Noise is something that is always there. Death, taxes and noise.

OT: Open Time, the observing time on a telescope that anyone can apply for – in contrast to the Guaranteed Time, the GT (see item), which is reserved for the instrument builders. The TAC (see item) chooses which observing proposals to accept.

PACS: One of the three instruments on *Herschel*. The letters stand for Photodetector Array Camera and Spectrometer, which is of no interest at all even to astronomers.

PI: Principal Investigator, the leader of a science project.

SAG: Specialist Astronomy Group. Matt Griffin, the Principal Investigator of SPIRE (see item), set up six of these groups of astronomers to plan the observing programme that would use the team's Guaranteed Time.

Signal: See item on noise.

Spectrometer: See item on spectroscopy.

Spectroscopy: The technique in which a special instrument, a spectrometer, is used to separate the blend of wavelengths that make up the electromagnetic radiation from a galaxy or other object. The deblended light – a spectrum – often contains spectral lines, dark or bright lines that are each produced by a specific chemical element or compound. I can use the spectral lines in a galaxy's spectrum to investigate its chemical composition, estimate its distance and measure its mass.

Spectra: See item on spectroscopy.

Spectral lines: See item on spectroscopy.

SPIRE: One of the three instruments on *Herschel*. The letters stand for Spectral and Photometric Imaging Receiver, which even those who built it often can't remember (possibly because the acronym doesn't make sense without using one of the e's in the last word).

Source: A location in the sky which is a source of some type of electromagnetic radiation. We talk about radio sources, submillimetre sources, infrared sources, etc. We rarely talk about optical sources, because if we have detected optical radiation it generally means we have a picture of the object, in which case we would usually call it a galaxy or star, or whatever it is, rather than a source.

Sub-millimetre (Sub-mm for short) waveband: Electromagnetic waves with wavelengths from 100 micrometres up to 1 millimetre (hence the name).

TAC: Time Allocation Committee – the committee of astronomers who decide which observing proposals for a telescope will be accepted.

Visible light: Electromagnetic radiation with wavelengths between 0.3 and 0.9 micrometres, the range that can be detected with the human eye (also known as optical radiation).

Visualism: A word I have made up to describe the prejudice that optical radiation must tell us more about the universe than radiation in other wavebands simply because our eyes happen to be sensitive to it

CHAPTER 1

Launch

MATT GRIFFIN ALWAYS PLANS his April Fool's jokes with great care. Every year on 1st April as I was getting ready for work, Keirsten warned me not to get taken in by Matt this time. And every year after I had cycled into work I would be thinking of something else, and I walked right into it.

By the end of March 2009, Matt had been setting me up for months.

I had been an astronomer for over 20 years, using telescopes all over the world, but when Matt moved to Cardiff a few years before this, he had asked me to join the team for the *Herschel Space Observatory*. This was the first time I had been involved with a space telescope and there were a lot of things I didn't like about it. It annoyed me that one-third of the observing time was reserved for the teams that built the instruments and their friends, although this now included me, and I didn't like the long timescales and endless delays.

For months I had been moaning to Matt at coffee that the launch of *Herschel* was sure to be delayed again. Matt always replied there wouldn't be another delay this time. It was costing the European Space Agency (ESA) too much to keep it on the ground. They had to fire it into space. Most of what I knew about space missions came from Matt, so I almost believed him.

A tall, softly spoken Irishman, Matt was the leader (the Principal Investigator or PI as we say in our world) of the team that had built one of the two cameras on *Herschel*. The name sounds very grand, but the leader of a big international space project has a lot of responsibility but not much real power. Matt had a team of 150 scientists in eight countries, but, except for a handful of scientists in Cardiff, he could never order anyone to do anything. He could ask a member of his team to do something, he could explain why it was necessary for the mission's success, he might try to cajole them with his quiet charm, but if the person ultimately refused there was not much he could do about it.

Over the previous decade, management challenges notwithstanding, he and his team had succeeded in building a state-of-the-art camera, which was now in Kourou, French Guiana, waiting to be installed on top of an Ariane 5 rocket. The team had been successful because they were all working towards a common goal, they trusted each other – many of them had worked together on previous projects – and, most importantly, everyone respected Matt.

One of Matt's favourite management techniques, whenever there was any contentious issue in the team, was to let everyone have their say, for as long as it took. After everyone had their say, and had heard everyone else speak, Matt would sum up and say it was clear the answer was A, even if some of us thought that most of us had just argued for B. Although most of us eventually realised what he was doing, his management-through-boredom technique worked because we all had a chance to give our two cents, we trusted him to make a sensible decision and we were often desperate for a drink. I once asked Matt how he handled the stress of being the PI. He did admit to sleepless nights, but I certainly never saw him anything but calm. Occasionally I would try to wind him up about something or other, but the most I ever managed to achieve was to see a look of mild irritation flicker across his face. Possibly his taste for April Fools' jokes was his way of dealing with the stress.

Matt's words about the delay made sense given everything he had told me about the telescope. The story of *Herschel* had always been about the money...

The 1970s had been a golden age for NASA. The early 70s saw the launches of *Pioneer 10* and *Pioneer 11*, which provided the first close-up views of the largest planets in the solar system: Jupiter and Saturn. Then in 1977 *Voyager 2* was launched, passing Jupiter in 1979, Saturn in 1981, Uranus in 1986 and Neptune in 1987. Until *Voyager* reached them, the two outermost planets, Uranus and Neptune, had only been points of light to telescopes on the ground. *Voyager* turned them into worlds. *Voyager* also presented for planetary scientists a smorgasbord of pleasures in the moons of the planets. The four largest moons of Jupiter were just points of light when they were discovered by Galileo, and they had remained points of light to every telescope ever since. *Voyager* revealed a dizzying diversity of properties, from Io, whose sulphur-belching surface made it look like a pizza, to Europa, the shiny moon, whose ice-covered surface is riven by a network of cracks caused by Jupiter's titanic gravitational field. The final years of the decade saw the development of the *Hubble Space Telescope*, which with its ability to fill in the fine detail of our blurred terrestrial view of astronomical objects has provided an epochal change in the human perception of the universe.

But on the other side of the Atlantic it was the dark ages. While NASA was setting out on the Grand Tour of the Solar System, ESA launched *COS-B*, *GEO-S* and *ISEE-B*, the first one a gamma-ray telescope and the last two spacecraft designed to investigate the Earth's magnetic field – all worthy space missions but jumbles of letters that nobody outside the astronomy community has ever heard of.

The big problem in Europe was money. When the European Space Research Organisation was reorganised in 1971 (it became ESA in 1975) the budget for space science was set at a level about one-tenth of NASA's and that budget had been frozen ever since. This created all sorts of problems.

Since there was so little money, when any did become available for a space mission, the tribes within the astronomy community fought ferociously to get their proposal adopted. Because it was only possible to fund one mission at a time, most of the community remained permanently disgruntled (in the words of one insider, the ESA science programme proceeded 'coup by coup'). There was also the additional problem that by the end of the decade the missions were becoming more ambitious. In the 60s, astronomers had been satisfied with relatively cheap missions, but now they were dreaming of major observatories containing many separate instruments, such as *Hubble*, or even missions consisting of more than one spacecraft. These were expensive.

The reason for the frozen budget was a hard political reality. Although ESA has no connection to the European Union, many of the political problems are the same. ESA is run by the ESA Council, which has a representative from each of the 22 countries that are ESA members, and these countries all have their own selfish national interests. In the ESA constitution, one of the rules is that it is only possible to increase the budget for space science if all the countries agree. For most of the 70s, it had not been possible to get unanimous agreement, and the budget stayed frozen.

There are compelling reasons why it is always a good idea to invest in space science. Space astronomy and planetary exploration expand our understanding of the universe we live in. Big investments in pure blue-sky science often lead, many years in the future, to some completely unexpected technological developments with huge significance for society – the worldwide web is the obvious example.… But forget all about that! It's the economy, stupid![1]

One hundred million euros invested in a space mission is not 100 million euros blasted into space. It is all spent on the ground. The money has been paid to high-tech companies around Europe to develop complicated pieces of space kit. Much of that money has been paid by these companies to smaller companies to build bits-and-pieces for this space kit. And much of the money paid to these companies has then gone to the highly skilled employees of the companies, who then pay for things in the wider economy that real people need: food, accommodation, cinema tickets, bicycles, cats, etc. There is a term that economists use – the multiplier factor – that stands for the number of euros that ultimately flows into the economy from an initial one-euro investment. The multiplier factor for investments in space science is a large one, but it is actually much better than that. Those technical developments mean that there is not just a one-off flow of money into the economy; there is a cascade of money into the future economy. This high-tech industry also requires people to be trained to a very high level, and these highly skilled people then go and do other things in the economy apart from space science – again an investment in the future.

Many of the European politicians in the 70s knew this. So why did they not line up to throw money at space science? The problem was that the language of politics is priorities.* There are always more things on a list of the things that need to be done than the money available; politicians need to get re-elected; and the arguments for space missions are hard to sum up in snappy political slogans. During the 70s, investing in space *did* bubble to the

---

* 'The language of priorities is the religion of Socialism' – Aneurin Bevan.

top of the priority list for some of the ESA members, but the unanimity rule meant that the budget stayed frozen.

In 1979, egged on by the astronomy community, the ESA executive tried to persuade the ESA Council to increase the budget by an ambitious 50%. They tried the additional argument of embarrassment. Since the combined economies of the European countries were as large as that of the USA and larger than that of the Soviet Union, it was embarrassing that Europe had such a pitiful space programme. Their proposal was considered at a meeting of the ESA Council on 28–29 June 1979. There are full meetings of the ESA Council at which the top politicians are present, at which budgetary decisions can be taken, and there are lesser meetings at which national delegations are present but the top dogs are not there. This was only one of the lesser meetings, but the national delegations made it clear that there was little-to-no chance that their governments would agree to increase the space-science budget.

Some of the smaller nations were in favour of an increase, but two of the biggest nations were completely against it. France had its own strong national space programme and had always been ambivalent about the ESA space-science programme; it preferred to spend any available money on its own programme rather than the ESA one. The UK was at the end of a long miserable decade for its national economy and had just elected a prime minister committed to retrenching the government budget. It is probably no coincidence that two months after Margaret Thatcher had been elected – a politician who claimed that, 'Pennies do not come from heaven. They need to be earned here on Earth' – the UK national delegation stated that their government could not possibly contemplate any increase in the space-science budget until the existing programme had been made more efficient.

The budget stayed frozen.

In the early 80s somebody tried again to break the ESA ice floe. Roger-Maurice Bonnet, who became Director of Science at ESA in 1983, was seen as a safe, conventional bureaucratic appointment. He had served on key ESA committees in the 70s and had spent his career at the elite Institut d'Astrophysique Spatiale in Paris, the heart of the French space-science establishment. He was not an obvious person to rip up the furniture, although it may be significant that he obtained his PhD in Paris in 1968, the year when student riots swept the French capital and the institutions of the French state almost collapsed.

On taking up his job, Bonnet's first conclusion was obvious enough to anyone who had been around ESA for such a long time. The big new observatories about which everyone was dreaming were just not possible without a long-term financial commitment. One problem was that all these dream observatories had missing pieces, bits of technology that did not yet exist. Everyone assumed it would be possible to develop these missing pieces, but this would require time and a sustained financial commitment; a commercial company will not take on the job of developing a key piece of technology, hire people, buy machines, if it thinks there is a chance the money tap may be turned off.

Bonnet's first action was highly unusual for a European bureaucrat, not known for their addiction to risk. The basic rule for bureaucrats is to accept the budget set by the politicians – they are the bosses, after all – and devise a program that fits within it. Bonnet had a different idea. Devise a program that blows the budget. Then dare the politicians to provide the money.

In November 1983, Bonnet issued a call for 'mission concepts' to the community. ESA received 68 different proposals, some not much more than preliminary concepts, some carefully worked-out proposals. They covered all aspects of space science. In May 1984, a meeting took place on the island of San Giorgio Maggiore in Venice, chaired by the Dutch astronomer Johan Bleeker, to consider the proposals. The meeting was probably the most important gathering in the history of European space science. Among the colonnades, towers and sixteenth-century splendour of the San Georgio Monastery – Europe in the past – the scientists came up with a plan for European space science in the future, which they called 'Horizon 2000'.

The plan was that Horizon 2000 would last 20 to 25 years, which meant there would be plenty of time to develop the missing technology. There would be four major missions covering the main areas of European space science, from exploration of the solar system to observations of the distant universe. Apart from these four 'Cornerstone' missions, the plan also contained some smaller missions. There would also be collaborative projects with NASA, one of which was *Hubble*.

But there was no money!

Bonnet's plan was simply to present the programme to the politicians and hope that the ambition and scope of Horizon 2000 would convince them to come up with the money. Initially, even within ESA people thought this was a daft idea, which awkwardly included Bonnet's boss, the new Director General, Reimar Lust. However, at a legendary dinner at a restaurant in Paris close to the Place de l'Etoile,[2] Bonnet and Bleeker convinced him this was a sensible approach. After all, nothing else had worked.

Bonnet's gamble was helped by the fact that, for the first time, the astronomy community cared about ESA. The breadth of the programme meant there was something in Horizon 2000 for everyone. In the 70s, an infrared astronomer might not care about ESA because the only project under development was an X-ray telescope. Now every part of the community had a stake in the programme. European industrialists liked it because most of the money would ultimately come to them. The last six months of 1984 were a period of intense lobbying. Industrialists lobbied ministers. Individual scientists wrote letters to their science ministers. In the offices and corridors of European capitals, scientists and delegations from ESA talked to civil servants, the civil servants talked to their ministers. One of the holdouts last time had been the UK Prime Minister, Margaret Thatcher. Bonnet wrote a personal letter to her, recalling a visit she had made to the European Space Research and Technology Centre (ESTEC) and her interest in the Giotto mission to Halley's Comet. (It probably helped that she had been trained as a chemist.)

Reimar Lust presented Horizon 2000 to a full meeting of the ESA Council, this time a full ministerial meeting, in January 1985 in Rome. The meeting agreed, unanimously, to fund Horizon 2000, increasing the budget by 5% each year for the next five years.

With Horizon 2000 and its Cornerstones, ESA had, for the first time, a programme to rival NASA's. The first Cornerstone would actually be two missions: one to study the Sun and one to study the Earth's magnetic field. The second would be an X-ray telescope, which would observe the mysterious black holes and quasars and the hot gas that earlier X-ray telescopes had discovered in clusters of galaxies. It was not clear which of the other two

Cornerstones would be launched first, but one would visit a comet or an asteroid. Its goal would be to land for the first time on one of these tiny objects, objects which have been in deep freeze since the beginning of the solar system. The other would be the mission that would become *Herschel*.

There was still one problem, and it was the same one as before: money.

Space projects are notorious for budget overruns because the technical challenges cost more to overcome than expected and take longer (which also costs money); new unexpected technical challenges materialise during the project; and over every space project hangs the unpredictable costs of space-proofing everything so that the instruments survive the launch and can exist in the radiation-soaked environment of space. These budget overruns are often jaw-droppingly large. The original budget allocations for the NASA-led *Hubble* and *James Webb Space Telescope* were already over one billion dollars, but both observatories ended up costing many times more than this. In the US, this is not usually a problem. Once the soft rain of space money starts fertilising the economies of their states, there always seems to be a host of congressmen and senators ready to defend the NASA programme. In Europe, though, the budget was always *the* problem. There was a fixed amount of money for the entire Horizon 2000 programme, and the unanimity rule meant that it wasn't going to change. The nice balance of the programme across the astronomy community also became part of the problem because if one part of the programme went over budget, it could conceivably cause the cancellation of another part of the programme, sending that part of the community back into disgruntlement with ESA. Exacerbating everything was that ESA had originally asked the member states for an annual budget increase of 7% a year. The ESA Council had given them 5% a year, which was enough to pay for Horizon 2000, just! As a result of all this, in the 1990s the Horizon 2000 programme ricocheted from financial crisis to financial crisis.

For *Herschel*, the emblematic crisis was the crisis over *Planck*. *Planck* was not one of the Cornerstones but a smaller (but still expensive) mission designed to study the cosmic microwave background, radiation emitted by the universe shortly after the big bang. In the mid-90s, as the result of a financial crisis caused by a series of unexpected events, including two missions lost shortly after launch, both *Herschel* and *Planck* came under threat of cancellation. Both missions managed to survive, as the result of behind-the-scenes manoeuvring and a letter-writing campaign from European scientists, including Steven Hawking and several Nobel laureates, but there was still a huge need to save money. To many parts of the community, especially the powerful tribe of planetary astronomers who would blow the whole ESA budget on missions to explore the solar system if anyone let them, *Herschel* and *Planck* looked pretty much the same. They would observe similar radiation, use much of the same cryogenic technology, and their scientific goals didn't really look that different to scientists interested in exploring the surface of Mars. There was a huge push to consider amalgamating the two missions. ESA did what all big organisations do when faced with a problem like this. It set up a committee.

The committee considered three options.[3] The first was simply to leave the missions alone, which would obviously save no money at all. The second was to keep the missions separate but to launch the spacecraft on the same rocket. The committee estimated this would save 80 million euros. The third was to merge the two missions and make a single observatory. The committee calculated this would save ESA 147 million euros.

Everyone in both the *Herschel* and *Planck* teams hated the third option. Their scientific goals were really very different and so they had very different designs. Trying to design a single spacecraft that could achieve the goals of both missions would inevitably produce an ugly hybrid – a kludge – something that would involve many compromises and might not even work.

Both teams shouted at ESA. Hearing the anguished screams from all sides and weighing them against the budget problem, ESA went with option two. Nobody liked it much. It made the schedule more difficult because both spacecraft needed to be ready for launch at the same time. It also increased the possibility of failure because, on top of the launch, there would be a risky moment when the spacecraft separated. But it did save some money – and, for both teams, it was a million times better than option three.

I hadn't been part of the project at the time. I hadn't even heard of *Herschel* back then. Most of what I knew about this crisis and all the other problems came from Matt and the other scientists in Matt's team. According to Walter Gear, who had been in Matt's team from the beginning, there was a simple reason for the decision to put both spacecraft on the same rocket: 'Putting them on the same rocket was a high-risk fudge to get them both done'.

Herschel and *Planck* did survive. And, despite all the problems with the budget, so did Horizon 2000. In 2009, 25 years after it was sketched on a piece of paper in Venice, the programme had largely done what it said it would. The first two Cornerstones had been launched and were now successfully in operation. The third Cornerstone, the mission to the comet that was now called *Rosetta*, had been launched and was on its ten-year journey to Comet 67P/Churyumov–Gerasimenko, where in 2014 its lander, *Philae*, would make the first-ever landing on one of the solar system's small objects. Several other smaller missions had also been launched, including *Mars Express*, which had produced spectacular high-resolution images of the surface of Mars. All that was left was the launch of the fourth Cornerstone: *Herschel*.

By March 2009 many of the people in the *Herschel* teams around the world were in a strange mood. *Herschel* itself had left ESTEC[†] in the Netherlands two months before. It had been taken by road to Amsterdam's Schiphol Airport and then transported by Antonov cargo plane, the only plane big enough to carry the three-tonne observatory and all its equipment, to the launch site in French Guiana. Now it was in the hands of the ESA staff at Kourou and many of the people in the *Herschel* teams had nothing to do.

People in the *Herschel* science teams like me, who would carry out the observing programmes with the telescope, actually hadn't had much to do for over a year, since the end of the intense period in which we had planned our projects and written our observing proposals.

[†]   European Space Research and Technology Centre

The difference, now, was that *Herschel* was on the move. When I write an observing proposal all I need to know about the telescope is a list of technical specifications: the size of its primary mirror, the wavelengths at which it will observe, the sensitivity of its cameras, the number of pixels in each camera, etc. It is all pretty abstract. Now, for me, *Herschel* had finally become real. There was a very real physical machine waiting to be launched into space in a hangar in Korou. Events were happening on the other side of the ocean that would lead inexorably to launch.

For the people in the teams that had built the cameras and the other instruments, *Herschel* had never been an abstract thing. They had been working on their instruments for almost a decade. An instrument they had handled with their hands in the lab was now in an observatory that would soon be launched into space. But their instrument was now out of reach, beyond any last-minute testing and tweaking. All they could do was wait and hope it survived the juddering journey into space, was turned on successfully and at least roughly met the design specifications that had been laid down a decade before. And to hope, if anything did go wrong, that it wouldn't be their part of the instrument that failed.

Carole Tucker, a postdoc in Cardiff at the time, had made the filters for Matt's camera. The camera was designed to observe sub-millimetre radiation. Sub-millimetre (sub-mm) radiation – like radio waves, X-rays, infrared radiation, microwaves, gamma-rays, ultraviolet radiation and even light itself (Figure 1.1) – consists of electromagnetic waves with a particular range of wavelength. The sub-mm waveband stretches from 100 micrometres, roughly the thickness of a human hair, up to 1 millimetre (hence the name 'sub-mm'). Carole and her boss, Peter Ade, a professor who seemed to live in the basement of the School of Physics and Astronomy at Cardiff (I never saw him enter or leave the building), own one of the weirder monopolies in the global economy. They are the only people in the world who know the secret of making good sub-mm filters. Over the last four decades, they have made the filters for virtually every sub-mm telescope, on the ground or in space.

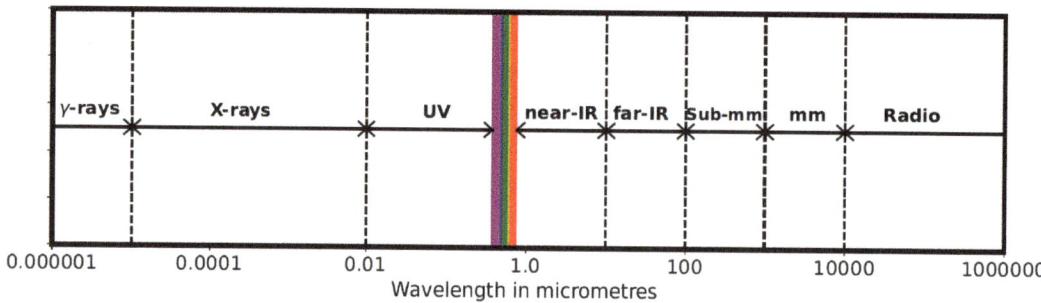

FIGURE 1.1 The electromagnetic spectrum. Visible light, the multi-coloured band in the centre, consists of electromagnetic waves with wavelengths between about 0.3 and 0.9 micrometres (one micrometre is one-millionth of a metre). The figure shows the other 'wavebands', wavelength ranges, in the electromagnetic spectrum. Many of the names are probably familiar, although the wavebands are human inventions and the wavelengths that separate the wavebands are arbitrary. Everything is electromagnetic radiation.

It is a surprising monopoly because most of the steps for making a filter are all there in the textbooks. The purpose of a filter is simple: let through radiation with the correct range of wavelengths for your camera and cut out all the rest. Making filters for sub-mm cameras, however, is particularly challenging because the sub-mm radiation from stars and galaxies is incredibly weak, which means it is crucial to cut out all the radiation at other wavelengths, which would otherwise make the faint sub-mm radiation impossible to detect. Peter and Carole's filters are unbelievably good, cutting out all this contaminating radiation to 1 part in 10^{15}, a 1 followed by 15 zeros.

The first steps in their recipe are the ones that are in the textbooks. First, make very accurate square grids of copper wires, in which the spacing of the wires in each grid has been chosen so it only lets through radiation in a certain range of wavelength. Then stack the grids together (upto 200), interleaved with polymer layers, with the grids and the spacing of the layers calculated to let through radiation in the range of wavelengths required for the filter. The final step is to melt the polymer layers to bind the grids together. Apparently, the magic (the group guards its secrets like a medieval craft guild) is in the bonding – just enough heat and pressure to bond the copper grids without distorting them.

As the maestros of sub-mm filters, they had naturally been given the job of making the filters for the *Herschel* camera. They had done many jobs like this before so it should have been routine; the filters for the camera were a little bigger than the ones they had made before but that was all.

In 2005, four years before launch, they hit a snag. When Carole tested some of the filters in the lab, she found that the radiation they let through did not have the correct wavelengths. The problem dragged on for months, with Carole making, testing and discarding hundreds of filters. It began to get more serious as the filter problem approached the so-called 'critical path', the moment when their little problem would be the limiting factor for the entire project, delaying the camera and possibly the launch of the observatory itself. Carole and Peter began to work 12-hour days. Everything else was put aside. More ideas were tried. More filters were made, tested and discarded. As a quasi-engineer, Carole couldn't admit to any stress – engineers solve problems, they don't get emotional about them – but as the months went past the weight of a billion-euro project began to descend on her shoulders.

Eventually, after almost six months, they solved the problem. It was actually a problem with the testing gear. The filters had been fine all along. Everyone in the instrument teams has similar stories: a decade of hard, steady work interspersed with manic periods as deadlines (or the dreaded critical path) approached – long days and lost weekends. But now, apart from a few people who were out at Kourou checking their instruments had survived the trip, Carole and everyone in the *Herschel* instrument teams had nothing to do. Göran Pilbratt, the *Herschel* Project Scientist, was still working hard. For the last 18 months, he had worked every day, including holidays and weekends, except for the occasional week when he turned off his mobile phone and went skiing or scuba-diving. But all everyone else in the *Herschel* teams was doing was waiting.

Six weeks to launch.

On 12th February, the spacecraft arrived at Rochambeau Airport in French Guiana. From there it was transported to Kourou in a special container flushed continuously with dry nitrogen to guard against any outside contamination. Once the big box was safely in a cleanroom in Kourou, the ESA technicians unpacked the spacecraft and checked it out.

A few days later, on 15th February, the telescope's mirrors were cleaned in preparation for the flight. The primary mirror, 3.5 metres in diameter, roughly one-and-a-half times bigger than the diameter of *Hubble's* mirror, was *Herschel's* eye on the universe. The telescope had been designed so that radiation from some distant galaxy would be reflected by the primary mirror up towards a small secondary mirror suspended above the primary. This radiation would then bounce off the secondary mirror, back towards the primary mirror, pass through a hole in its centre and be detected by the telescope's instruments in the cryogenic chamber below. Cleaning the mirrors was a delicate job. The ESA technicians, dressed in cleanroom suits and paper hats and overshoes, which made them look like scene-of-crime technicians, worked in pairs, each checking the other did not accidentally touch the mirror's surface. They cleaned the mirrors by spraying them with carbon dioxide 'snow', which vaporised almost instantaneously as it landed on the mirror.

In early March, members of the instrument teams checked that *Herschel's* three instruments had survived the journey and were working. Everything checked out. The three instruments were 'good to go'.

On 27th March, the spacecraft, once again in a big box, was moved to the S5 building, five kilometres further towards the Ariane 5 waiting to take it into space.

The S5 building is where spacecraft fuelling happens. Although most of the oomph for sending *Herschel* into space would be provided by the huge Ariane 5 rocket, after which the spacecraft would coast for a month until it reached its destination four times as far as the Moon, it also had its own tiny rockets for orbital adjustments. The rocket fuel du jour is hydrazine. Hydrazine is highly toxic and dangerously unstable, which is why it makes good rocket fuel. The S5 building at Kourou is well away from the other buildings for a reason.

On the last day of March, I was working at home on my laptop late in the evening. I checked my mail one last time before going to bed. There was a message from Matt to our entire team. He was passing on, in confidence, a memo from a meeting at ESA headquarters in Paris that he had attended earlier that day.

I looked at the list of attendees. There were only a couple of names I recognised. Matt had been there and also Göran Pilbratt, the *Herschel* Project Scientist. There were a couple of people from the ESA Space Operations Centre, which would be responsible for operating the observatory once it was launched, a representative from *Planck* and a couple of names from ESA headquarters. It was a standard ESA memo, full of acronyms, written in grammatical English but with a few awkward constructions that showed the writer was not a native English speaker.

There was a problem with *Herschel*. The memo stated that the technical checkout at Kourou had revealed a major problem with the *Herschel* Attitude and Orbit Control System. This was a problem because this is the system used to point a telescope at a target; a telescope that can't point is, well, pointless. It was a big problem, but it was not a catastrophe because it had been caught before launch. It would mean an annoying delay, but the telescope could simply be shipped back to the Netherlands, repaired and then returned to Kourou. The meeting in Paris had been convened to come up with a solution. They had looked at this option and decided that it would mean a delay of 6–9 months and cost 100–120 million euros, which was obviously a lot but only 10% of the cost of the mission. The meeting had also looked at a second option. As I remembered what Matt had been telling me recently about the history of the mission, I could feel my stomach knotting as I read on.

They had found the old industrial study carried out in the 90s that looked at the possibility of merging the two missions. They had realised that the two spacecraft could be bolted together and operated as a single observatory. The memo said that there were attachment points on both spacecraft, which meant that this could be done quite easily. The *Planck* pointing system was working fine. If the two spacecraft were bolted together, the *Herschel* observations could be done in the first three years, with *Planck* being used to point *Herschel* at its targets. Then with the two spacecraft still clamped together and now spinning, *Planck* would take over and carry out its own survey of the sky. This option would only cost 20 million euros and would require no delay.

The meeting recommended the second option to the ESA Director General.

I was surprised but I wasn't surprised. I had been expecting something like this from the bureaucrats. It was almost a relief the disaster had finally happened. Of course, it wasn't really a disaster. The mission would still go ahead, and it might work. At least I wasn't a member of the *Planck* team who would have to wait three years for their mission to even start. But with such a stupid solution something was sure to go wrong. All for a saving of just 10% of the total budget! Why did they have to ruin such a beautiful mission?

Well, that was that! It had happened. I fired off a reply to Matt, copied to all 150 members of the team, venting my feelings about the idiocies of the ESA bureaucracy, who knew the cost of everything but the value of nothing, and went to bed.

When I got to work the following morning, I realised I had been had yet again. I wasn't alone. About half of us had fallen for it. We had missed a couple of clues. Two of the names of the ESA bureaucrats at the meeting in Paris were Karl Munchausen and Jacques Poissondavril.‡

At least it made a break from the waiting.

In the first week of April the fuelling began. For its three-year mission *Herschel* needed 256 kilogrammes of hydrazine, roughly seven times the mass of the petrol in my car. Spacecraft fuelling is done slowly, carefully, with all electrical equipment turned off and covered with protective silver foil. Dressed in protective 'scape suits' with their own air supply, and looking much like the Apollo astronauts, pairs of ESA technicians fuelled the spacecraft over two days in five-hour shifts. The fuelling finished, everyone left the cleanroom. Someone somewhere flipped a switch. The spacecraft power came back on. Nothing happened. Good to go.

11th April 2009. Five weeks to launch.

Back in Cardiff, I was beginning to find it hard to concentrate. I still had lectures to give but otherwise I had plenty of time for research. *Herschel*, though, was going to be so

---

‡ A Poisson d'avril is a joke made in France on April 1st. Karl Munchausen, Baron Munchausen, a real person in eighteenth-century Germany famous for his tall tales, was turned into a fictional character who told even taller tales; the Baron was not amused.

revolutionary, if it worked, that all my current research projects now seemed stale. The launch of *Herschel* meant that we were going to be able to start using a new electromagnetic waveband for astronomy. Astronomy in the visible waveband (optical astronomy) has been done forever. Radio astronomy began when the scientists who developed radar during the second world war came back and starting using the wartime technology to build radio-telescopes. And astronomy in the wavebands in which the atmosphere is not transparent (Figure 1.1) began with the launch of the first space telescopes in the 60s and 70s. Every time in the past when astronomers had the opportunity to make observations in a new waveband, they discovered something new. The astronomers who were lucky enough to be there at the beginning of radio astronomy discovered quasars, pulsars, radio galaxies and the cosmic microwave background radiation. The astronomers who were there at the beginning of X-ray astronomy discovered that in clusters of galaxies there is invisible gas between the galaxies with a temperature of 100 million degrees. None of these discoveries would have been made if we had stuck to the visible waveband.

And now we, too, would have the same opportunity. Apart from a few narrow bands of wavelength in which the atmosphere is partially transparent, which can be used from the ground in exceptionally dry weather, observations over most of the sub-mm waveband are only possible from space. Apart from a brief foray across this new electromagnetic frontier a few years earlier with a balloon experiment, this was virgin territory. Nobody knew what was out there. We would be there at the beginning.

We were also fairly sure that this new waveband held the answers to two huge questions. For over 200 years astronomers have been trying to understand how the stars and galaxies were born. The thing that has always stood in our way is something that is completely invisible.

An interstellar dust grain is an insignificant thing, only about one thousandth of a millimetre in size and much too small to see with the naked eye, but dust grains together are a big problem. Dust grains, tiny solid particles, are found everywhere is space, and it is their ubiquity that makes them such a problem. The grains scatter and absorb light, which means the dust acts as a 'smoke' reddening and concealing many of the objects in space. This is especially true of stars and galaxies right at the beginning of their lives because the dust in the clouds of gas in which they are born makes traditional optical telescopes virtually useless. By the time of the launch of *Herschel*, the astronomers who use telescopes that observe in the visible waveband had made little progress, even with telescopes as powerful as *Hubble*, in answering these questions.

We were confident, though, that we would be able to answer them because with a sub-mm telescope we would have a way of 'seeing' through the dust. When a dust grain absorbs some light, it is heated by the energy in the light. Not a lot. The energy in the light heats the dust to only a few tens of degrees above absolute zero, but it is still enough for the dust grain to emit sub-mm radiation. A new-born star or galaxy might be hidden completely by the dust from an optical telescope, but we would be able to observe them by detecting the sub-mm radiation from the dust. With *Herschel* we would have the chance to find answers for two of the biggest questions in astronomy.

It is not surprising I was finding it hard to concentrate.

During this period, I spent a lot of time bored, prolonging coffee breaks and wandering around the department trying to find people to talk to. I ended up talking a lot to my own research group (they had to talk to me).

Simon Dye, one of my two postdocs, had worked with me for years. We complemented each other well. I would have some idea for a project and try and do it myself, which in astronomy usually involves writing some computer program. After a week I'd give up in frustration. Simon would take over and after a few months would present me with some beautiful piece of analysis. Simon has the craftsman's skill that was used to make the medieval cathedrals – he just uses it to construct elegant pieces of software. He also has an admirable array of practical skills, is a meticulous DIYer and even delivered a child on the verge of a motorway when someone went into premature labour. Come the apocalypse and the collapse of civilisation, Simon would be a good person to have around (although, thinking about it, he would probably regard me as too old and useless and kill me for meat).

I didn't know Robbie Auld so well. We had just hired him as a postdoc to work on the data we hoped to get from *Herschel*. I didn't know how good Robbie would be as a scientist, but his constant jokes and buoyant personality were already lifting everyone's mood; Robbie was the kind of person every office needs.

I was not sure what would happen to Robbie if there was an incident with *Herschel* ('incident' is a comforting word suggesting that if only an ESA 'tiger team' with the correct skillset is assembled, an appropriate solution will be found, not the towering inferno that would ensue if the Ariane 5 exploded on launch). Robbie hadn't asked me, and I had been too afraid to check. The rumour was that the last time an astronomy mission exploded, all the contracts were cancelled from the moment of the explosion.

The baby of the group was Matt Smith. In anticipation of the deluge of data that would come from *Herschel* (if everything went well), we had taken him on as a PhD student the previous summer after he finished a physics degree in Cambridge.

On 20th April, the magical transformation of the helium in the tank below the telescope's primary mirror began.

The big challenge of sub-mm astronomy is that everything in the universe emits sub-mm radiation. Interstellar dust is so cold it emits most of its radiation in the sub-mm waveband (Chapter 2). Although the Sun and the Earth emit most of their radiation at much shorter wavelengths, they are so luminous that they are also also intense beacons at sub-mm wavelengths. A telescope, even its mirror, and all the instruments on the telescope also emit sub-mm radiation. Unless something is done about it, the flood of sub-mm radiation from everything else swamps the faint sub-mm signals from stars and galaxies.

ESA planned to solve the problem of the Earth and Sun's radiation by sending *Herschel* to L2, the second of the five Lagrangian points.[§] Once it was launched, *Herschel* would still need to travel almost one million miles, four times as far as the Moon, until it entered an orbit around L2. Once at L2, the Earth and the Sun, from the perspective of the telescope,

---

§   The five Lagrangian points are the locations at which a small object can orbit in a fixed pattern with two larger objects. In the Earth-Sun system, the second Lagrangian point is on the opposite side of the Earth from the Sun.

would always lie in roughly the same direction,[※] which would make it much easier to position the telescope's huge sunshields to keep the flood of sub-mm radiation from them away from the telescope's mirrors.

But this wouldn't solve the problem of the sub-mm radiation from the telescope itself. The cooler an object, the less radiation it emits. The only way anyone has come up with for solving the problem of the flood of sub-mm radiation from every part of a sub-mm telescope is to cool everything down to a ridiculously low temperature, which is one reason why my kind of astronomy has lagged so far behind every other kind.

The coldest temperature possible is absolute zero, which is −273 degrees Celsius or 0 kelvins on the temperature scale used by astronomers.[※※] The universe is not quite as chilly as this because of the cosmic microwave background radiation, which fills the whole of space and keeps everything in the universe at a toasty 2.7 kelvins. But to detect the sub-mm radiation from distant galaxies, the cameras on *Herschel* would need to be even colder than this.

The ESA engineers had put a tank of liquid helium on *Herschel* to keep its instruments cold. The tank was at the bottom of a huge vacuum flask – the cryostat – which was just below the primary mirror. The instruments were also inside the cryostat, on the top of the tank, sealed off from the rest of the world by the lid on the cryostat, which would be opened once the telescope was in space. The helium, 365 kilogrammes of it, was at a temperature of 4 kelvins. But this was not cold enough.

On 20th April, the ESA technicians started to 'pump out' the tank. The pumping created a vacuum above the liquid, causing some of the liquid to boil away. As this happened, the remaining liquid became colder. When the temperature decreased to 2.17 kelvins, the helium was transformed into a superfluid, a strange state in which the atoms of a liquid flow past each other without any friction. The technicians topped up the tank and kept on pumping, topped up some more and pumped again…. The temperature eventually stabilised at 1.65 kelvins, with the helium now all converted from standard helium I into superfluid helium II. The tank of liquid helium in Building S5 at the Centre Spatial Guyanais in Kourou was now one the coldest places in the universe.

On 29th April, *Herschel* was put back in its container and slowly moved 10 km to the Bâtiment d'Assemblage Final (BAF – the Final Assembly Building).

Two weeks to launch.

In Cardiff, as the launch got closer, we stopped talking at coffee about the projects we would do once *Herschel* was in orbit. Suddenly everyone was an expert on the safety record of different types of rockets. Ken Wood, the sales director of the spin-out company that sold Peter and Carole's filters, a Mancunian with a salesman's easy humour, attempted to reassure me: 'Don't worry, Steve, the Ariane 5 has a 90% success rate'. He didn't succeed. I wasn't sure whether Ken was trying to wind me up or not, but I could do the maths. A 90% success rate means a 10% failure rate. That didn't seem that small. I had worried about a lot

---

[※] An object at L2 takes the same time to travels around the Sun as the Earth, which means for an object at L2 the Earth and the Sun always lie in the same direction.

[※※] Temperature in kelvins is equal to temperature in degrees Celsius + 273

of things in my career as an astronomer – proposals, grants, travel arrangements, instrumental failure, the weather – but I had never had to worry about my telescope blowing up.

It didn't help that the fate of one of the previous Cornerstones was on everyone's mind. *Cluster*, part of the first Cornerstone, was designed to study the Earth's magnetic field, which is important because it shields us from high-energy particles from the Sun. *Cluster* had been due to fly on the second of the new Ariane 5 rockets. But when the commercial satellite due to be carried on the first Ariane 5 fell through, ESA offered the *Cluster* team the berth at a reduced price. New rockets always use some of the same technology as previous rockets, and there was some software on the Ariane 5 left over from the less powerful Ariane 4. On 4th June 1996, 37 seconds after launch, some of this leftover code, 'dead code' that was not meant to be used, misinterpreted the greater acceleration of the more powerful rocket and switched off its inertial navigation system. In the most expensive software bug in history, the Ariane 5 veered off course and was destroyed by the rocket's self-destruct system.[††] (There's no such thing as a free launch.)

Successful launches are rarely on the news, unsuccessful ones always are. I remember seeing the explosion of *Cluster* on the evening news in the 90s. The news item showed a group of *Cluster* scientists sitting around a table, watching the launch on a big TV screen on the wall. A buffet lunch had been arranged to celebrate the launch and there were plates of sandwiches on the table. The camera caught the expressionless faces of the scientists as they watched the rocket explode. At the time, I didn't know any of the *Cluster* team and I hadn't cared that much. I had even found it funny, possibly it was the plates of sandwiches. Now I was having some late-blooming empathy; I could feel their pain as they watched their mission blow up in front of them. (It can't have been much of a free lunch either.)

Deadlines concentrate the mind. As launch day approached, I realised there were a lot of things I didn't care about. I didn't really care about Robbie's contract – he could always get another job. I didn't care about the decade that Matt Griffin and the other instrument builders had spent making their instruments – they knew the risks. And I didn't even care about the time I had wasted myself on *Herschel* or most of the observing projects that would not be carried out if *Herschel* exploded on launch. What I *did* care about was one particular project.

Astronomers measure out their lives by observing proposals. The objects in the winter sky are different from those in the summer sky, so for most telescopes there are usually two observing cycles six months apart. For telescopes where this doesn't matter – space telescopes like *Hubble* and the *James Webb Space Telescope* or telescopes on the ground that observe 24 hours a day like the Atacama Large Millimeter Array (ALMA) – the observing cycle is chosen for human convenience and usually lasts about a year. As I write, we are now in Cycle 25 for *Hubble* but only Cycle 6 for the much newer ALMA. For any telescope, the observing cycle starts with the announcement of opportunity – the AO. The AO tells astronomers around the world (astronomy is an inclusive pursuit – most telescopes are open to anyone from any country, professional astronomer or not) that there is the opportunity to submit a proposal for observing time on the telescope. The astronomers submit

---

[††] The Belgian scientists who had designed the self-destruct system were apparently happy that it had worked so well.

their proposals. The Time Allocation Committee, the TAC, reads the proposals, accepts some and rejects most.

I like to think of a telescope proposal as a miniature art form, a haiku or sonnet perhaps, but possibly it is more like a short story of fixed length. The rules are very strict, and the TAC will not even read proposals that infringe on them even a bit. For a medium-sized telescope, the rules are typically one page about the scientific results you plan to get, one page with the details of how you will use the telescope's instruments to get these results, and two pages for figures and references (no cheating, there is a minimum font size). Within that small space, the story you write needs to be convincing enough that the TAC will rank your proposal, for popular telescopes like *Hubble* or ALMA, in the top 10% of the thousand-or-so proposals that have been submitted, which are the only ones that will get some observing time on the telescope.

To write a convincing story, there are some useful guidelines. A TAC is composed of astronomers like you, but they will not be exactly like you. In my research, I try to understand the origin and evolution of galaxies. A TAC might contain one person who works in this research field, but most members of the TAC will be from wildly different research fields, perhaps even a planetary astronomer who will probably know nothing about galaxies at all. It is therefore important to write your story not for specialists but for astronomers in general. It is also important to remember that the members of the TAC are not paid to do this; they have day jobs and they are often reading the proposals in their spare moments when they are not at their sharpest (common places for me: on a plane, early in the morning before work, after work in the evening, occasionally after a couple of pints of beer, sometimes even in the bath).

If science was like the claims of some philosophers, writing a telescope proposal would be easy. The Austrian philosopher Karl Popper claimed that scientific progress was produced by the insights of great theorists (he was thinking of Einstein, I suspect), with the contribution of experimenters or observers like me being confined to testing the hypotheses and predictions of the theorists. I have never liked this idea since it makes my role a secondary one; nobody gets out of bed to test a hypothesis, unless it is a very important one like Einstein's prediction that light is deflected by gravity, which I suspect may be how Popper got the idea. If Popper is right, writing a telescope proposal would be easy. All you would have to do is describe the hypothesis to the TAC, explain why it is important, and how you are going to test it.

Real observing proposals are not like that. To get an observing proposal accepted, you need to find some way of exciting the TAC – to persuade them that the proposed observations are sure to lead to something new and interesting, even if they won't really. I may need some last few observations to finish off some project, which are almost certainly not going to change the project's main results. I still need to find some way of making the proposed observations exciting to the TAC. And all in one page and in words even a planetary astronomer can understand.

We cast our finely written stories on the waters and the TAC stomps on most of them in its wellington boots. Approximately three out of every four proposals for most telescopes are rejected, and for popular telescopes like *Hubble* or ALMA it is more like nine out of ten. (Gmail sends all e-mails from ALMA to my spam box, and since most are rejections it must know something.)

But at the moment I submit a proposal, I have always convinced at least one person. The day of a telescope deadline is always chaos. In ALMA Cycle 6, 1838 proposals were submitted, and roughly 80% of these were submitted in the 24 hours before the deadline. In the last few hours before a deadline, proposal writers incorporate last-minute comments, fiddle with figures to squeeze them into the space, cut out inessential words to save those last three sentences marooned beyond the page limit, and if possible – it rarely is – read the other proposals on which they are authors. By the time I have submitted my own proposal, at the end of all this frenzy, it feels like I have been at some kind of revivalist meeting. I have managed to convince myself, at least, even though when I started the proposal I wasn't sure the idea was any good.

The moment I push the submission button, the frenzy starts to lift. A few months later, when the TAC delivers its decision, I am usually nonchalant about a rejection – and sometimes even grateful when I had already realised it was a daft idea and the project would be a waste of time. But occasionally, very occasionally, I am waiting for the decision with dread. I know the project is so good that any rejection will be a travesty of justice. And since the TAC is merely a jury of my intensely fallible peers, I know a travesty of justice is perfectly possible.

There have only ever been a few projects like this for me. One of them was the project I wanted to do with *Herschel*.

The idea for the project came from a conversation in my office in Cardiff in 1999 between me and my then PhD student, Loretta Dunne. We thought the idea was amazingly cool, but we also knew the project was impossible with the primitive sub-mm cameras available at the time. We thought about the project again sometime in the early 2000s when we heard about *Herschel*. But when we did some calculations to check whether the project was possible, we concluded reluctantly that the telescope's cameras would still not be sensitive enough.

In 2007, suddenly everything changed. We were at a meeting to discuss large projects that might be carried out with *Herschel*, which was being held at ESTEC in the Netherlands. Shortly before the meeting, we had heard a new, much better, set of figures for the sensitivity of the main *Herschel* camera, which meant the project was now possible. At the meeting, we met other people who were interested in a similar project. We assembled a team. I put together the proposal, using Loretta's and my original idea plus the ideas of our new collaborators. We asked for 1,200 hours of observing time, an outrageous request since this was 10% of all the time available and most proposals were only asking for 100–200 hours. We argued that the huge amount of time was merited since the project would produce a revolution in several research fields. We crossed our fingers.

When the TAC announced its decision, it said there was no way we could have 1,200 hours of observing time. They said the complicated scheduling necessary for *Herschel* (some parts of the sky would only be visible for brief periods six months apart) meant it would be impossible to carry out such a large programme before the big tank of helium ran out. They gave us 600 hours. We were very happy. We had padded our time request because we knew that TACs always try to cut down on the total time request – 600 hours would be enough.

We had the time in the bag. *Herschel* would be in space so there wasn't the usual gamble with the weather. There was just the little matter of the launch.

On 29th April, *Herschel* arrived at the BAF. The Ariane 5, the size of a 14-story building and one of the most powerful rockets ever built, was already there (Figure 1.2).

Once in the BAF, a crane lifted *Herschel* to a high bay level with the top of the rocket. An Ariane 5 can launch two satellites at once using a metal cage structure. *Herschel's* launch buddy, the *Planck* spacecraft, was already in its place in the cage, its nose just visible (Figure 1.3). Using a crane, the ESA technicians guided *Herschel* into its position

FIGURE 1.2 The Ariane 5 with *Herschel* on board in the Bâtiment d'Assemblage Final (BAF – the Final Assembly Building).

Credit: ESA

FIGURE 1.3   An artist's impression of the layout at the top of the Ariane 5. *Herschel* is on top with *Planck* in the cage underneath. The two space companions are covered by the metal fairing that formed the tip of the rocket.

Credit: ESA

above *Planck*. Once it was in position, they bolted everything together, installed *Herschel's* electrical connections and checked everything was working. Good to go.

During the next ten days, the main activity was more filling of the cryostat to replace the evaporating helium.

On 10th May, the ESA staff closed the cryostat, disconnected the helium-filling equipment, powered down the spacecraft and removed its umbilical links to the ground. The tip of the rocket, a white metal fairing, itself as tall as a four-story building, was then hoisted above the cage structure. As the fairing was lowered, the technicians removed the silver foil that had protected the telescope, giving them a brief view of the mirror. Then *Herschel* vanished from human view for the final time.

Four days to launch.

On 12th May, the doors of the BAF opened. The Ariane 5 began its very short but infinitesimally slow railway journey to the launch pad (Figure 1.4)

One day to launch.

14th May – Launch Day. My alarm goes off at six. Half a second, maybe even less and everything comes flooding back. This is a day when everything will have changed by the evening, one way or the other. An either-or day. Exam days, medical appointments, and now launch days.

The drumbeats of fate may be beginning but normal life has to carry on. I go downstairs and have some coffee, have a shower and eat breakfast on the deck. I kiss Keirsten goodbye and cycle off to work. It is a beautiful spring day, fluffy white clouds blowing steadily from the west across the pristine blue sky. As I cycle along the river past the cathedral, the water falling over the weir sparkles in the sunshine. In Bute Park, the big lime trees along

FIGURE 1.4 The Ariane 5 with *Herschel* on board on the launch pad. Good to go!

Credit: ESA/CNES/Arianespace – Photo Optique Video CSG, J.M. Guillon

the river are finally in leaf. The water tumbles along between the trees, playing in the sun. The birds are singing. Here in the city, I could be deep in the countryside. It is one of those days when Wales seems the most beautiful country in the world.

When I get into work nobody else seems to be hearing the drums. There is the usual bustle in the department office. It is after term but there are still a few students wandering the corridors. Lecturers hurry in and out of the mailroom. At the end of the corridor, I catch sight of Robbie having a laugh.

When I get to my office, I try to do my usual routine – check my email, look at the list of papers that have appeared in the archive overnight – but my heart isn't in it. Mentally I go through the schedule for the day. Since Matt is in Kourou, I am one of the senior people left in Cardiff, so I am one of those responsible for everything going well. If the launch itself goes wrong, I may even have to say some inspirational words to pull everyone together (I can't think what I could possibly say but hopefully something will occur). The launch party will start at 12.00. There will be four talks, one introducing *Herschel* and then three talks describing the science it will do. I will give the second science talk. Then we will watch the launch. At some point Rhodri Morgan, the famously disorganised First Minister of Wales, will arrive and talk to us, we assume before the launch.

Since I am feeling antsy, I decide to go down and check on the room. When I get there it is empty. The technical staff haven't set it up yet. But they are a competent lot and I am sure it will be ready. It was a bit daft even coming down here. I go back to my office and go through my talk for the fifth time. I go down to coffee. We sit around not discussing the launch, talking about this and that; it seems bad karma now to talk about launch statistics.

Finally, at 11.50 it seems acceptable to go down to the room. There's still nobody there, but I am one of the ones responsible for the day, after all. The technical staff have done their stuff. There's an array of wine bottles, beer bottles and plates of sandwiches covered in cling-film. As with PhD vivas, this preparation seems strangely premature – what do we do with all these provisions if something goes wrong? People – students, postdocs, academics, technical staff – start to filter into the room. Nobody seems nervous. There is just the usual hum of university folk anticipating a free lunch and this time free booze. People start to chat, drink and eat sandwiches. One of our outreach people prowls the room with a camera, interviewing people for posterity. I say something inspirational about the importance of interstellar dust.

The talks begin. Peter Hargrave, a postdoc who has spent ten years working on the camera, gives the first talk, introducing *Herschel*. On one of the screens behind him, we can see the live feed from Korou showing the Ariane 5 on the launch pad and the banks of people and terminals in Mission Control. The sound has been turned down so we can hear Peter but the activity on the screen is distracting. My friend Derek Ward-Thompson, one of the other professors, gives the next talk. Derek talks about how *Herschel* will be used to observe the birth of stars. I don't really listen. I am thinking about my own talk and I am watching what is happening in Korou. Rhodri Morgan hasn't arrived yet.

I am up next. My talk goes by in a daze. I talk about how we will use *Herschel* to observe galaxies, but I can almost feel in my back the activity on the screen behind me. After me,

Phil Mauskopf, another professor, talks about how *Herschel's* fellow passenger on the Ariane 5, *Planck*, will be used for cosmology. Rhodri Morgan still hasn't arrived.

The talks over, the sound on the live feed from Korou is turned on again. It's all in French so it's a little difficult to be sure what's going on, but the activity on the screen starts to look more purposeful. The chat in our room dies down. As we listen to the voices of the Mission-Control team, the screen starts to show the Ariane 5 on the launch pad. I sneak a look at the audience. Everyone is now staring raptly at the screen like a crowd watching a penalty at a football match. The seconds tick away and eventually the French Mission-Control director announces that the countdown is about to start.

Dix, neuf, huit, sept, six, cinq, quatre…. The gantry moves away from the rocket. *Trois, deux, un….* The main rocket engine of the Ariane 5 flames into intermittent and then steady life. Clouds of smoke billow away from the rocket. The rocket laboriously leaves the Earth, rising slowly above the lush foliage on twin pillars of flame from the solid rocket boosters. Everyone cheers. Somebody shouts (I think it is Ken Wood), 'Don't come back'.

At 16 seconds after launch the rocket vanishes into the clouds. A few seconds later it emerges, moving at a slight angle to the vertical, the by-now single pillar of flame trailing a roiling river of smoke. At 138 seconds after launch, the solid rocket boosters fall away, the rocket continuing to climb in the spotless blue sky at an ever-increasing angle to the vertical, now powered by the main engine. We cheer again. Ten minutes after launch, at a height of about 145 km, well outside the thin skin of the Earth's atmosphere, this engine and the fuel tank drop away. The rocket continues its journey into space on the power of the third-stage engine.

Twenty-six minutes after launch, something happens that I don't realise has been a worry. The third stage has now fallen away, and the spacecraft is continuing on its way without power. Somebody says that *Planck* and *Herschel* should now have separated. I don't realise this is a big deal, but it is. Launching *Planck* and *Herschel* as a single payload – putting all the eggs in one basket – had been one of the money-fuelled compromises that were made in the 90s. Nobody knows yet whether separation has been successful. We won't know until Mission Control establishes radio contact with the spacecraft. The tension builds up again. After 12 minutes we hear that the ground station at Malindi in Kenya has contacted *Herschel*. The status of the spacecraft is 'nominal' which in space-speak means OK. Good to go!

Rhodri Morgan finally turns up. He gives an off-the-cuff 15-minute speech perfectly designed to push all our emotional buttons. After Rhodri has spoken, we feel even better about *Herschel*, what we've done, and what we're going to do next. Rhodri and his entourage sweep out, on their way to his next gig to inspire another group of people. We start the celebrations.

This is not the last worry. Thirty days after launch, the lid on the cryostat needs to open – and if the lid doesn't open that's the end of the mission. Then the instruments will be switched on. And who knows whether these have survived the launch. Some of these instruments have moving parts and if one of these fails the whole instrument is dead – even the scientists who built them are crossing their fingers about these. Matt has told me that *Herschel* with its instruments is the most complex machine ever been sent into space.

None of this is reassuring. There is also the worry that there is something wrong with the telescope's optics, which seems ridiculous except that this is exactly what happened in the 90s with *Hubble*, when after launch its mirror was discovered to have the wrong shape. If something similar happens with *Herschel*, we won't be able to send astronauts up to fix it. This was possible with *Hubble* because it is in a low-Earth orbit but *Herschel* will be too far away.

But all this is for another day.

CHAPTER 2

Beginnings

WHEN I WAS A kid there was no such thing as a sub-mm astronomer. Even when I was a PhD student in the early 80s, there were no real sub-mm astronomers. What there were, were a bunch of physicists who liked a big challenge. As technological challenges, it would be hard to find a much bigger one. Trying to detect sub-mm radiation coming from space from the surface of the Earth is virtually impossible. The water vapour in the atmosphere absorbs most of the sub-mm radiation from space and there are only a few narrow bands of wavelength in which any radiation gets through at all. These 'atmospheric windows', which exist in only a very few, very dry locations – the summit of Mauna Kea in Hawaii, Antarctica, the Atacama Desert in Chile and that is to name almost all of them – are notoriously flaky and can vanish for weeks at a time. Then there is the overwhelming flood of photons from everything else – the atmosphere, the Earth, the telescope itself and the detectors – that swamps the infinitesimal signals from space, which means that the filters to cut out unwanted radiation have to be almost impossibly perfect and the detectors have to be cooled to a smidgeon of a degree above absolute zero.

On top of these, there was the not insignificant problem in the early 80s that there were no actual sub-mm telescopes. Since the mirror on a regular optical or infrared telescope is perfectly fine for reflecting sub-mm radiation, it was possible in principle to do sub-mm astronomy by attaching a sub-mm detector to one of these telescopes. The time allocation committees, though, were naturally very reluctant to grant observing time to do this, since it was perfectly obvious to everyone that sub-mm astronomy was so primitive that it would not tell you much about planets, stars or galaxies or really anything at all. But here the sub-mm physicists caught a small break: it is possible to do sub-mm astronomy during the daytime for a few hours after dawn. The time allocation committees were prepared to give them some of this time to try out their instruments since nobody more important wanted it.

For the sub-mm gang, an observing run was therefore a complicated business. First, get your instrument working in the lab. Then crate up the instrument and all the cryogenic paraphernalia and fly it over to Hawaii (the usual destination in the 70s and 80s) and take it up the mountain. Get it working as well on the summit as it was back in the lab (another big

 DOI: 10.1201/9781003195290-2

challenge) without getting in the way of all the other astronomers and engineers working in the telescope dome. Then spend a frenetic one to two weeks going up the mountain to observe when everyone else on the summit is going down for breakfast. Then do everything in reverse: pack up your kit, take it down the mountain and back to your home lab, make some improvements to the instrument and start again. By the mid-80s, they had detected some planets, a few clouds of dust and gas in the Galaxy, and a very few other galaxies. Since there are approximately 6,000 stars visible to the naked eye, it is not hyperbole to say that in the early 80s our knowledge of the sub-mm sky was a lot worse than the view of the sky accessible to any Australopithecus a couple of million years earlier.

No real astronomers took any of this very seriously. The sub-mm gang had overcome some huge technical challenges and shown that sub-mm astronomy was possible, but they had never discovered anything interesting. To observers like me, they were just the irritating people who turned up at dawn when we were all tired and wanted to go to bed. In the 80s, not surprisingly, there were not many places where it was possible to do sub-mm astronomy at all. The main centres of expertise in the UK were Preston Polytechnic and Queen Mary College in London, not one of the fashionable London colleges but out in the sticks in the pre-gentrified East End. Back then, sub-mm astronomy was not a fashionable pursuit.

Sub-millimetre radiation, infrared radiation, gamma-rays, X-rays, radio waves, ultraviolet radiation – all in a sense are just light but outside the range of wavelengths in which our eyes are sensitive and called by a different name. Light in the conventional sense is something we take for granted, but without it we would have difficulty making our way around in the world. As I lie here in a hammock on the hottest day of the year so far, the photons from the Sun, particles of sunlight, bouncing off my body are the only reason I can see my toes.

The sunlight is a stream of particles, but in a confusing way, not understood at an intuitive level even by the quantum physicists who study them, the particles are also waves.* The wavelength of any wave, from light photons to the waves on the ocean, is the distance between two peaks or two troughs in the wave. There are the objective properties of light like the wavelength, which we can build instruments to measure, and there is the subjective way our brains perceive the light. We *see* the light waves with the shortest wavelengths as blue and the light waves with the longest wavelengths as red, with the wavelengths in

---

* This is called the 'wave-particle duality' by quantum physicists, but the term is a little misleading. Photons and other sub-atomic particles – electrons, protons, neutrons and so on – do sometimes behave like particles and do sometimes behave like waves, but the truth is they are neither. On a scale one million billion times smaller than the objects around us, these sub-atomic 'things' do sometimes behave like the concepts we have from our human-scale world – snooker balls, waves etc. But the truth is, they are something else, something that is beyond the direct apprehension of beings in a world one million billion times larger.

between giving us all the colours of the rainbow. The reason the Sun appears yellow to me, if I sneak a look at it now, is the blend of wavelengths in the sunlight. And the reason that my toes appear pink when I look at them – or more accurately now I look at them properly, a grubby off-white – is that my skin has absorbed light of certain wavelengths out of the sunlight, so the blend of wavelengths that is reflected from my toes is no longer the same blend that fell on them. The reflected photons are transformed – after they have travelled to my eye, passed through my pupil, been focused by the lens and converted by the cells in the retina into electrical signals, which pass along an optic nerve and are processed by the algorithms encoded in my brain – into the cluster of neural signals I identify as 'off-white'. As a scientific phenomenon, the simple act of seeing is therefore quite wonderful. It embraces just about everything in science: nuclear physics (because that's how the Sun produces its energy), optics, quantum physics, biochemistry, neuroscience and psychology. There is also of course the everyday wonder of watching the light play on the surface of the ocean, or of watching the insects travel from flower to flower on a hot summer day.

In the objective scientific world, light consists of electromagnetic waves with wavelengths between 0.3 and 0.7 micrometres, a micrometre being one-millionth of a metre, roughly one-third of the thickness of a filament in a spider's web. All the other names in the list above are electromagnetic waves with wavelengths outside the visible range, each being the name (the names are arbitrary human inventions) of the electromagnetic waves in a particular wavelength range (Figure 1.1). Any kind of wave – a sound wave or a water wave, for example – travels through a medium. The medium of an electromagnetic wave, the electromagnetic field, is not something, like air and water, that we can sense directly, but although we can't sense it with any of our five senses, we know it is there, even in a vacuum, from its effects on equipment that measures electrical or magnetic forces.

The first person who realised there was something more to light than meets the eye was William Herschel, an astronomer, who in 1800 carried out one of the classic experiments in science. Unlike modern astronomers, with all their fancy high-tech equipment, Herschel simply looked through a telescope and sketched what he saw. When he looked at the Sun, so that he could do it safely, he used coloured filters to block out most of its light. But when he did this, he noticed he could still feel the Sun's warmth on his face. He also noticed that it felt warmer through a red filter than a blue filter.[1] It would have been easy for him not to follow this up – he was busy with many astronomy projects – but one of the reasons he was a superb scientist was his ability to spot the things which were worth spending time on. This one he decided to investigate.

One of the reasons the experiment is a classic is its simplicity. He positioned a glass prism so that it intercepted the sunlight passing through a south-facing window (Figure 2.1). By bending the sunlight with different wavelengths by different amounts, the prism transformed the sunlight into a spectrum, a rainbow of colours – in the same way that a rainbow is produced by water droplets in the atmosphere. Herschel positioned a table so that the spectrum fell on its surface. His only other experimental apparatus was three thermometers. He used one to test whether there was a temperature rise when he placed its tip in the different bands of light. The other two showed his care as an experimental scientist. He placed them on either side of the spectrum to check that any temperature rise on the first thermometer was caused by the coloured light and not by something else.

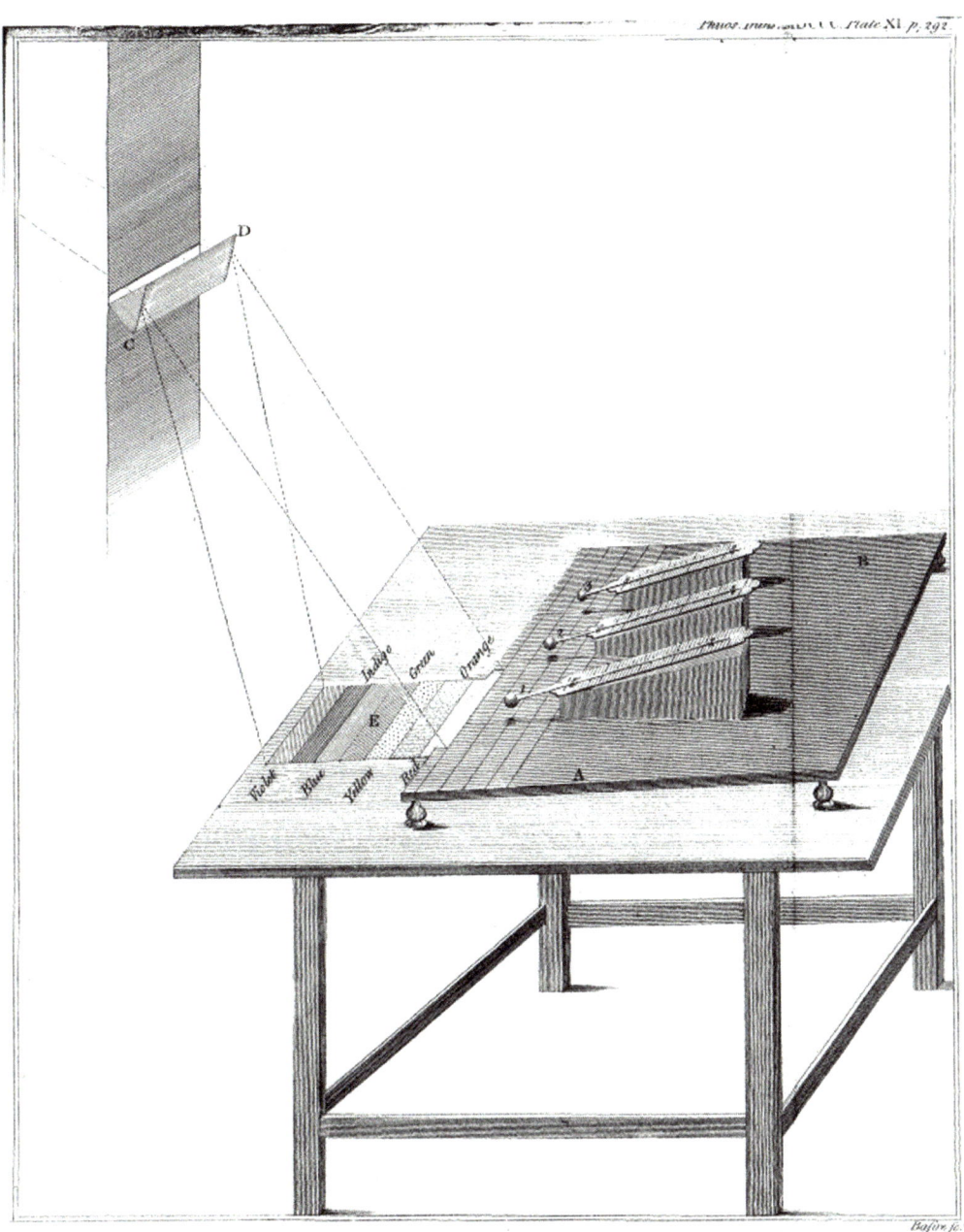

FIGURE 2.1 The apparatus William Herschel used to discover infrared radiation. The sunlight came in through the window on the left and passed through a prism, with the spectrum of light falling on the table. The three thermometers are shown to the right of the spectrum.

The temperature on the first thermometer did rise when its tip was in one of the coloured bands, and he also found that the rise in temperature was greatest when the thermometer was placed in the red part of the spectrum. Since the temperature increased steadily from one end of the spectrum to the other, he conjectured there might be a second type of radiation, separate from the light and invisible, which was carrying the Sun's heat. He speculated that its effect might extend beyond the visible spectrum.

This was easy enough to test.

When he placed the tip of the thermometer beyond the spectrum's red end, where there was no light visible at all, the temperature still went up. He had discovered *infrared radiation*.[†] In the language of the twenty-first century, he had discovered that the electromagnetic spectrum extends beyond the visible range.[‡]

The reason I can see my toes is that they are reflecting sunlight, but everything in the garden around me, everything in the universe, also *emits* radiation.

The blend of wavelengths in an object's radiation depends on its temperature. This relationship is summarised in a law of physics[§] which states that the wavelength at which the object's radiation is strongest is inversely proportional to its temperature, which means that if the object's temperature increases by a factor of five, for example, the wavelength at which the radiation is strongest decreases by a factor of five. The Sun's radiation is strongest in the visible waveband (Figure 2.2) and its temperature is about 6,000 kelvins.[¶] The temperature of everything around me today – the trees, the grass, my body, my hammock – is roughly 20 times lower than this, which means the radiation from all the objects in the garden is strongest at a wavelength 20 times greater than the Sun's radiation, which is in the infrared waveband (Figure 2.2).

Everything in the universe follows the same law. The gas between the galaxies in clusters is exceptionally hot, with a temperature of about 100 million kelvins, and therefore most of its radiation is emitted at very short wavelengths, in the X-ray waveband (Figure 2.2), which is why it was only discovered when the first X-ray telescopes were launched. At the other extreme, the temperature of interstellar dust is only a few tens of kelvins,

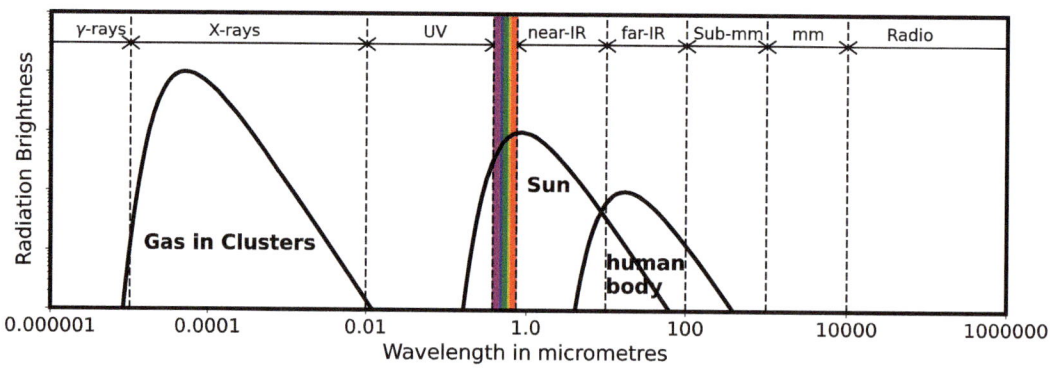

FIGURE 2.2   How the strength of the radiation emitted by an object depends on wavelength for three different objects: the gas in clusters of galaxies, the Sun and the human body.

[†]   Infrared because one meaning of 'infra' is beyond, so infrared – beyond the red.
[‡]   He thought he had discovered that the Sun emits two types of radiation. Back then, of course, he did not know about electromagnetic waves and thought that light consisted of a stream of particles.
[§]   Wiens Law.
[¶]   This is the temperature of the Sun's photosphere, which is effectively its surface. It is much hotter below the photosphere.

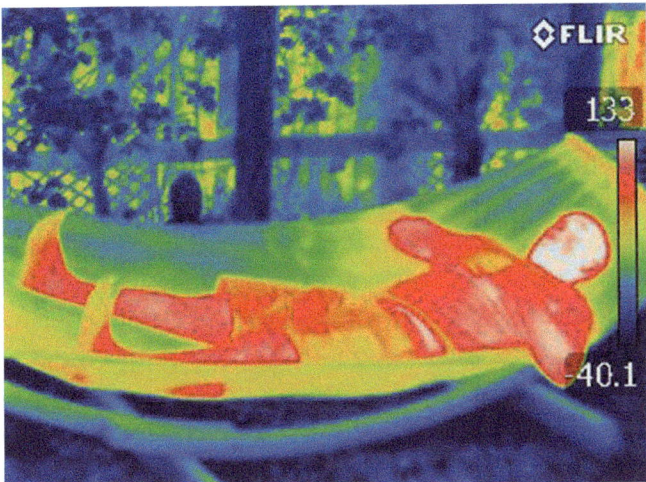

FIGURE 2.3   Infrared picture of the author, with the colours showing the strength of the infrared radiation: blue, green, yellow, red and white in order of increasing radiation strength.

and it therefore emits most of its radiation at very long wavelengths, in the far-infrared and sub-mm wavebands.

I can see the trees and everything in the garden today because of the sunlight they reflect – visible light. But if I had infrared eyes, rather than relying on the sunlight washing over the world, the world would look a very different place. A picture with an infrared camera only gives a hint of what it would be like to have infrared eyes because the picture needs to be transformed into a visual picture for us to be able to see it.

Figure 2.3 shows a picture taken with an infrared camera of me in my hammock. So that we can see the picture, the strength of the infrared radiation is represented by different colours. My body is shown in reds and whites because it is emitting more infrared radiation than all the objects around me. My head is emitting most because the brain accounts for so much of the body's energy consumption. My feet are emitting less infrared radiation than the rest of me because they are colder, as I have just been walking on the grass in my bare feet. If you look closely, you can see my toes.

Back in the 80s, I was at a fashionable institution. In 1981, I started a PhD in the radio-astronomy group in the Cavendish Laboratory at Cambridge University. Unlike Preston Poly and Queen Mary College, both founded in the twentieth century, Cambridge was founded almost a millennium ago in 1209. It is an institution steeped in history and full of beautiful medieval buildings, but it was only in 1874 when the Cavendish Laboratory was founded that the university started to become the scientific powerhouse it is today. The Cavendish was where two of the fundamental particles of nature, the electron and the neutron, were discovered; it was the place where Crick and Watson discovered the structure of DNA (which they announced at the nearby Eagle Pub by saying they had found the secret of life); and it was also the place in the years after the second world war where a new kind of astronomy began.

Stories from the aboriginal people of Australia show that we have been using our eyes to study the sky in the visible waveband for at least 10,000 years,** but the rest of the electromagnetic spectrum has been used for less than 100 years. The first of the electromagnetic frontiers was opened because of the developments in radio technology that happened during the war. The frontier burst open in many places, but in Cambridge the pivotal moment was when Martin Ryle, a young scientist who had worked on radar during the war, obtained a fellowship at the Cavendish Laboratory and decided to build a radio telescope.

In the early days, the Cavendish radio astronomers, directed by Martin Ryle,†† led the world. They invented many of the basic techniques of radio interferometry, in which many radio dishes are joined together electrically, effectively creating a single giant dish. They used these telescopes to make the first radio surveys of the sky, finding places in the sky that are sources of radio radiation ('radio sources', for short). Due to the steep learning curve with any new technique, the first two Cambridge catalogues of radio sources were riddled with errors, but the third Cambridge (3C) catalogue of radio sources[4] started a revolution in our understanding of the universe. It contains the position and brightness of the 328 brightest radio sources in the northern sky. Most of these were things we would not know existed without radio-telescopes – new types of beast.

Initially, nobody knew what to make of them. In the early days, astronomers who used regular telescopes, telescopes that observe in the visible waveband, took pictures to see what was there at the positions of the radio sources. Sometimes there was a galaxy there and other times a star, but it was not obvious why these galaxies and stars should be strong radio sources and not others. At the positions of some sources there seemed to be debris present and it was not clear what was emitting the radio waves. The first radio-telescopes showed hardly any detail, but better radio-telescopes became available, and radio-astronomers began to get detailed images – radio pictures. Theorists also gradually uncovered the physical processes that are producing the radio waves.

One of my favourite sources is 3C 405, the second brightest source in the catalogue, which is also known as Cygnus A because it is the brightest source in that constellation. Figure 2.4 shows pictures of Cygnus A in the radio waveband and the visible waveband. The picture in the visible waveband shows nothing much – a faint galaxy that we would have no idea is unusual without a radio-telescope. But the radio picture reveals one of the biggest beasts in the universe. The radio source from end to end is about 390,000 light-years and, since the pictures are on the same scale, it extends well outside the galaxy seen in the optical picture.

** Some of the aboriginal people maintain an oral knowledge of the night sky including the positions of all the visible stars.[2] One of their stories mentions a star close to the south celestial pole. The south celestial pole moves in a small circle in the sky because of the precession of the Earth's axis. The last time there was a star close to it was 12,000 years ago,[3] showing that the aboriginal people must have been studying the sky for at least this long.

†† He shared the Nobel Prize for Physics with another of the Cavendish radio astronomers, Anthony Hewish, in 1974.

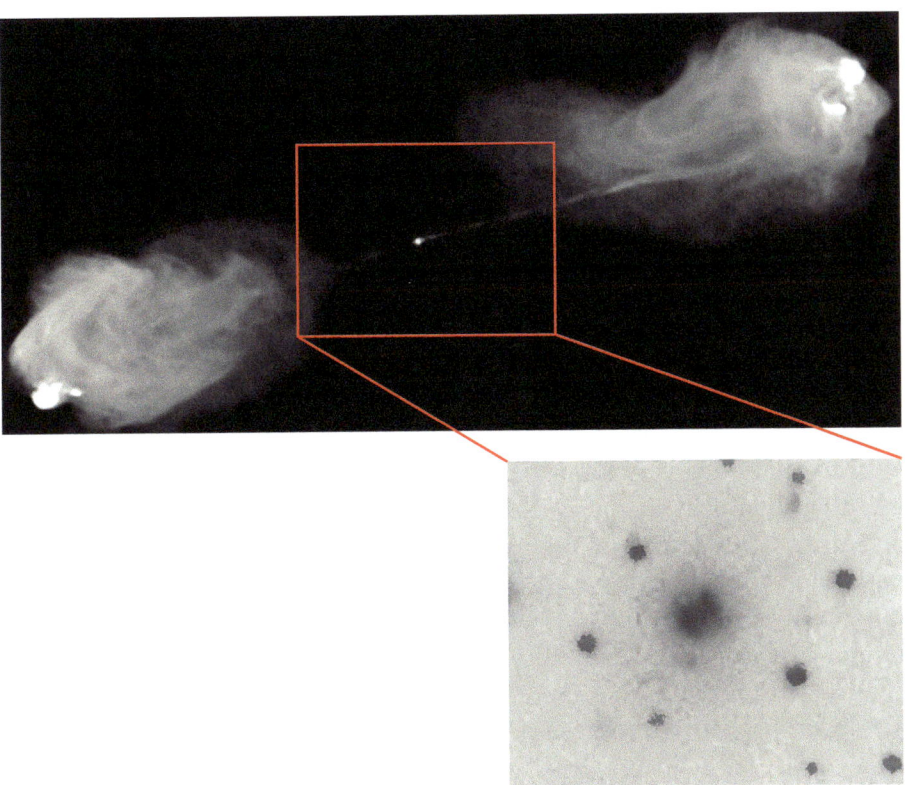

FIGURE 2.4   **Top** – radio picture of 3C 405 (Cygnus A). **Bottom** – picture in the visible waveband of the area covered by the red square in the radio picture (shown in black and white rather than as our eyes would see it).

Credit: NRAO, R. Perley

Astronomers gradually figured out what was going on. At the centre of the galaxy, there is a black hole with a mass of about one billion times the mass of the Sun. We can't see it – black holes don't emit radiation – but if gas starts to fall into the black hole energy is generated, and the release of energy produces two jets of ultra-fast sub-atomic particles. As a result of the collision between these jets and the gas surrounding the galaxy, electrons are accelerated to close to the speed of light. These electrons emit radio waves, producing the woofly[‡‡] structure in the radio picture. One of the lessons here is the importance of complementary views. The picture in the visible waveband shows the galaxy itself, the radio picture reveals the quick-as-lightning (almost) electrons accelerated in the collisions between the jets and the surrounding gas, which we would also be able to see if we also had an image taken with an X-ray telescope.

[‡‡] 'Woofly' was coined by Julia Riley, who was in the radio astronomy group at the same time as me. It does describe the gauzy, billowy, filamentary structure in the radio picture of Cygnus A rather well.

Cygnus A is a 'radio galaxy'. The stars that astronomers discovered at the positions of some of the 3C sources turned out not to be stars but 'quasi-stellar radio sources' – 'quasars' for short. A quasar is quite similar to a radio galaxy, a distant galaxy with a supermassive black hole at its centre. The difference is that in a quasar the energy produced as the gas swirls into the black hole generates a large amount of visible light – so much that it swamps the light from the surrounding galaxy, making it very hard to see.

Radio-telescopes also made it possible to look at old beasts in new ways. The brightest source in the 3C catalogue is 3C 461, Cassiopeia A, which is one of the sources where there is debris present in the picture in the visible waveband. This is the debris left from a super-nova, the explosion that occurs when a massive star collapses. Without radio-telescopes, we would never have known that among this debris there are ultra-fast electrons accelerated in collisions within the debris, which emit radio waves. (As it happens, without a radio-telescope we would never have known about the existence of this supernova. Astronomers have used the radio observations of 3C 461 to infer that the supernova explosion was sometime in the 1700s, but it was not noticed at the time, probably because the supernova was hidden by interstellar dust.)

The discoveries kept on coming. Two other Cavendish astronomers, Jocelyn Bell Burnell and Anthony Hewish, discovered radio sources that flash, sources where the radio waves seem to turn on and off.[5] At first they thought these pulsating sources (pulsars for short) might be signals from an extraterrestrial civilisation, but we now know these are neutron stars, the densest thing (apart from a black hole) in the universe – another new type of beast. Observations with another radio-telescope, not in Cambridge this time, led to the discovery of the cosmic microwave background radiation, evidence that the universe had a beginning. In the 50s and 60s, with new and better radio-telescopes being built all the time, it seemed the discoveries would go on forever.

As a student in the group in the 80s, it was exciting to be surrounded by such history. I would cycle out to the Lords Bridge Observatory to pick up the data from the day's observations with the One Mile Telescope. The telescope was impressive itself, three giant metal-meshed dishes supported by black metal frames, whose strangely unbalanced appearance made them look like giant robot figures striding across the countryside – an impression enhanced by one of the dishes being at a different spot every time I visited (Figure 2.5). But there was also the Half-Mile Telescope, four smaller dishes spaced along a railway line half-a-mile in length, and the newest telescope, the Five-Kilometre Telescope, eight dishes disappearing off into the haze of the flat Cambridgeshire countryside. The data from the One Mile Telescope was recorded by holes on paper tape (state-of-the-art technology from 1964). When I got back to the Cavendish to pass the paper tape through the paper-tape reader, it was inspiring to be surrounded by the people that had built these radio-telescopes. Martin Ryle was now a campaigner against nuclear weapons and was rarely seen in the department, but many of his students and associates, the pioneers on this new frontier, were still in the building.

By the 80s, however, the pace of discovery had slowed. The last of the Cambridge interferometers, the Five Kilometre Telescope, had been finished in 1972, and it had now been outclassed by the Very Large Array in New Mexico, which was completed in 1980.

FIGURE 2.5   One of the two fixed dishes of the One Mile Telescope. The other was moved along a track. One of my more embarrassing moments as a PhD student was when I drove the moveable dish, 120 tonnes of it, into one of the fixed dishes.

While I was there, the main goal of most of the members of the Cambridge group was to get a better understanding of radio galaxies like Cygnus A. There was general agreement that a galaxy like this contains a supermassive black hole at its centre, which is somehow generating jets of sub-atomic particles, but most of the details of how this happens were still unclear. In the four years I was in the group, these details were what most of the group were working on. I didn't get the impression they were making much progress.

By the end of my PhD, I knew I wanted to carry on in astronomy, but I was also a little jaded with radio astronomy. I had done a PhD in radio astronomy because that was the kind of astronomy that was done in Cambridge, but I was ready for something different.

At the time, I thought this was because I just wasn't that interested in the physics of ultra-fast electrons in intergalactic space. Stars and galaxies were interesting, but clouds of ultra-fast electrons somehow weren't. But now I think it was because radio astronomy was going through one of those dead times that every research field goes through. There are times when new techniques or ideas break a field wide open and there is an avalanche of discoveries. Radio astronomy was now in one of those other times. After the completion of the Very Large Array, there would be no large new radio interferometer for over three decades.[§§] This is not to say that there was no scientific progress with existing telescopes. It was just rather slow and incremental – and a little boring.

Fortunately, something had sparked my interest. My PhD supervisor John Baldwin rarely had time to meet me, but sometime at the beginning of my second year, I happened to see him in the corridor. The radio observations I had carried out in the first year had not yielded any interesting results. John, as usual on his way somewhere else, suggested I try something different. Why not observe some of the sources I was finding with an optical telescope? I took him seriously. I applied for time on a telescope at the Kitt Peak Observatory in Arizona. I was awarded the time, and John and I went out to do the observations.

With a mirror only 90 centimetres across, the Kitt Peak 0.9-metre telescope was hardly a big telescope, but it had recently acquired one of the very first CCD[¶¶] cameras. There is now one of these in every mobile phone, with many more pixels than in our camera, but it was a big deal at the time because it made it possible to use our little telescope to observe fainter objects than had been possible before. Our support astronomer showed us the new camera, told us it had cost 100,000 dollars, and although he didn't exactly tell us we would have to pay for any damage, the implication was-be very careful.

I decided to use our observing time to see whether radio galaxies like Cygnus A are located in galaxy clusters. I planned to use a very simple method: take a picture of each radio galaxy and count the number of galaxies around it.

Because the telescope was one of the smallest ones at the observatory, there was no telescope operator (TO), and so John and I had to do everything ourselves. We would point the telescope at one of our radio galaxies, start an exposure, and after 20 minutes look at the picture on a computer screen. We only saw a cluster of galaxies once – the project was pretty much a bust – but in every image there were always some galaxies. As I looked at them, I realised that each of these little smudges probably contained 100 billion stars. Each of these stars probably had planets around it.

This was much more exciting than clouds of ultra-fast electrons. One thing I had got wrong was the temperature. Even in Arizona, April nights on a mountain summit are very cold. But this only added to the romance. Enduring the cold on a mountain in Arizona, observing galaxies billions of light-years out in space, trying to keep awake, watching the sky gradually brighten at the end of the night. This was real astronomy.

The plan at the end of my PhD was therefore to give up on radio astronomy and apply for postdocs at places that did astronomy the old way. For almost a year the astronomy community did not seem to have heard of the plan and I got turned down for postdoc after

---

[§§] It would have been easy enough to build interferometers with more dishes but the computing power to process the data has not been available until recently. A spectacular new radio-telescope, the Square Kilometre Array, is now under construction in Australia and South Africa.

[¶¶] CCD stands for charge-coupled device.

postdoc, spending the last six months of my PhD supported by the UK welfare system, which was a lot more generous in the 80s than it is today. But eventually, with a lot of perseverance and a lot of luck, I did get offered a postdoc. After being turned down for about 30, I was eventually offered one at the place that at the beginning would have been my top choice: the Institute for Astronomy in Hawaii. I borrowed some money from my parents and flew off to Honolulu.

The Institute for Astronomy is at the entrance to Manoa Valley, only a mile from the beaches of Waikiki. On the beach the sun is usually shining, up in the valley it is often raining, and the Institute is in the middle, which meant that on many of the days during my three years in Hawaii I would look through my office window and see a rainbow arcing from one side of the valley to the other. I was lucky to spend three years in such a beautiful place.

But it was not the beauty of the place that had taken me there. The real lure for an observer is the Mauna Kea Observatory, which is not on Oahu like the Institute but on the 'Big Island', the Island of Hawaii. On the summit of Mauna Kea, a 14,000-foot quiescent[***] volcano, the observatory in the 80s was the best place in the world to do astronomy. For old-style optical astronomers, the attraction was the clarity of the images possible with the telescopes there – the crispest, most detailed pictures of the sky until the launch of *Hubble*. For other astronomers, the attraction was the dryness of the atmosphere above the observatory, which made it possible to observe the universe in some of the non-traditional wavebands.

And for postdocs like me, there was also the enticement of the amount of observing time on offer. Postdocs don't always find it easy getting observing time, but the Institute has a slice of the observing time on every telescope on the mountain, so even a postdoc there can often get more time than they can use. Once I realised this, I decided to spend the next three years trying to observe on as many different telescopes as possible.

The people I had come to work with at the Institute were Eric Becklin and Gareth Wynn-Williams, the odd couple of infrared astronomy. The infrared waveband (Figure 1.1) was one of the new frontiers, which had been opened up for astronomy in the 60s. Eric and Gareth had been there at the beginning. Eric had even had one of the first infrared sources named after him.[†††]

The two of them were opposites in every way. Eric I found a little scary. He was tall and blond with a Scandinavian background. I heard so many stories about him that I found it hard not to think of him as some kind of modern-day Viking rather than a professional astronomer. Short, dark, and reserved, Gareth's natural milieu was probably a book club. They were also opposites when it came to astronomy. When he went observing, Eric barked out commands to the TO, modifying the observing programme minute

---

[***] It erupted 4,500 years ago and according to the US Geological Survey it is expected to erupt again at some time in the future – 'extinct' would sound more reassuring than 'quiescent'.

[†††] The Becklin-Neugebauer object, an object in the Orion Nebula that is completely hidden by dust, almost certainly a newborn star

by minute, the only astronomer I've known who's done this – astronomy as jazz. Gareth usually fell asleep. His strengths lay elsewhere. He was good at all the post-observation stuff that Eric wasn't interested in: data analysis, writing papers and writing grant applications – the bread-and-butter of a research program. I worked for them, paid out of a grant they had from the US National Science Foundation. But they were both very generous. They let me do whatever research I wanted as long as they found it interesting.

Gareth was the one I worked with most. One day in his office he mentioned that he had heard about a new instrument that was coming to one of the telescopes on the mountain. It was a sub-mm instrument being built by a team in the UK, but it was not one they would keep taking back to their lab. It would stay in Hawaii for astronomers like us to use. Gareth suggested that we apply for some observing time.

After a little research in the library, I came up with an idea. I found a couple of papers by scientists who had predicted the temperature of interstellar dust. When an interstellar dust grain absorbs starlight, the energy in the starlight heats up the dust grain. If it had no way of losing energy, the dust grain would keep on getting hotter as it absorbs more starlight. But everything in the universe, dust grains included, emits radiation. The authors of the papers made the reasonable assumption that the rate at which the dust grain loses energy from the radiation it emits is balanced by the rate it gains energy from the absorbed starlight. They predicted that the temperature of interstellar dust is about 20 kelvins. I realised that, with this new instrument, we now had a way of testing their prediction.

As I explained earlier, the wavelength at which the radiation from an object is strongest depends on its temperature; if the temperature falls by a factor of two, the wavelength at which the radiation is strongest increases by the same factor and so on. I reckoned that if the papers were right about the temperature of the dust, the radiation from it should be strongest at roughly 100 micrometres, which is right on the boundary between two wavebands: the far-infrared waveband and the sub-mm waveband (Figure 1.1). The far-infrared waveband is one of the electromagnetic wavebands in which the atmosphere is completely opaque – nothing gets through at all. Sub-mm radiation is also almost completely blocked by the atmosphere, but there are a few narrow wavelength bands in which the atmosphere is partially transparent. It was these atmospheric 'windows' that the new instrument was going to use.

A few years before I moved to Hawaii, the first far-infrared observations of galaxies had been made by the Infrared Astronomy Satellite (IRAS). One of its big discoveries was that galaxies are surprisingly strong far-infrared sources, and the far-infrared radiation from a galaxy's dust is often as strong as the visible light from its stars. Figure 2.6 shows the IRAS measurements for a typical galaxy. IRAS observed each galaxy at four wavelengths, and this galaxy is typical in that the radiation is strongest at the longest wavelength: 100 micrometres. This wavelength is right on the boundary between the far-infrared and sub-mm wavebands, which raises the obvious question: Does the strength of the radiation keep on climbing into the sub-mm waveband?

This depends on the temperature of the dust. If the temperature of interstellar dust really is about 20 kelvins, as the papers I had found were predicting, the radiation would be strongest right on the boundary. If the dust is cooler than this, the radiation would be

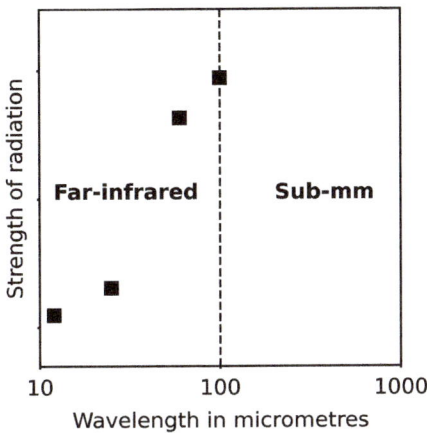

FIGURE 2.6 The strength of the far-infrared radiation measured by IRAS for a typical galaxy.

strongest somewhere in the sub-mm waveband. With the new instrument we would be able, for the first time, to measure the strength of the sub-mm radiation from galaxies, which should allow us to find out the wavelength at which the radiation from the dust is strongest, which would allow us to estimate its temperature.

I wrote an observing proposal, it was accepted, and a few months later Gareth and I flew over to Hilo on the Big Island to make the observations.

After landing in Hilo we drove up to Hale Pohaku, the residence where the astronomers stay when they are observing, which is 9,000 feet above sea level although still 5,000 feet below the summit. After dropping our bags and having dinner, Gareth drove us up to the summit to meet the support astronomer who would look after us during our observing run.

Today the route to the summit is a tarmacked road, but back in the 80s it was a rubble track, which zigzagged back and forth up the sides of the mountain. On that evening the sky was clear and on every zig Gareth had to slow because of the glare from the setting sun. Not far above Hale Pohaku all vegetation stops, and we drove through an arid volcanic landscape that makes me think of the surface of Mars. I don't really remember, but I assume that when we drove over the last brow of the mountain and the telescope came into view I got the same stressed feeling I always get when I see the telescope I am about to use.

Our telescope was the United Kingdom Infrared Telescope and our support astronomer was Bill Duncan, a tall moustachioed Englishman. After shaking hands with us, he took us into the dome to see our instrument. The instrument, with the very unromantic name of UKT14,[‡‡‡] was hanging below the telescope, a red cylinder among a tangle of tubes and wires. Meeting the instrument is something I have never quite understood but it is an accepted ritual of the astronomy community. I tried to look intelligent while Bill pointed out its finer features and to grunt appreciatively at all the right moments. Then we went back into the control room for the briefing about how to use it.

[‡‡‡] It was the 14th instrument to be installed on this telescope.

When I had asked Gareth before the observing run how many magnetic tapes[§§§] I should take to store our data, he had laughed. In the control room there was none of the shiny electronic paraphernalia and flashing screens of optical astronomy. We would be using only two simple devices. The first of these was the chart recorder, a ridiculously low-tech device in which a long strip of graph paper gradually moves from one spool to another, with a pen tracing a flickering red line along the centre of the strip. Gareth, an old-time infrared astronomer,[¶¶¶] got quite excited by this, but to me it looked amusingly primitive, the bad guess of a soviet sci-fi director from the 1920s of what the twenty-fifth century would be like.

Gareth tried to explain its purpose. One of the biggest challenges in sub-mm astronomy is the flood of sub-mm radiation from the atmosphere itself. With UKT14 we would be able to measure all the sub-mm radiation from within a circular aperture, a circle on the sky. We could try to measure the sub-mm radiation from a galaxy by the obvious way of moving the telescope so the galaxy fell within the instrument's aperture, but the big problem was that the sub-mm radiation from the galaxy would be swamped by the flood of sub-mm radiation from the bit of the Earth's atmosphere that also fell within the aperture.

The way we would overcome this, Gareth explained, was that the telescope would do a little dance. It would move a small distance back and forth across the sky every ten seconds, in the first position with the galaxy centred in the aperture and in the second position with no galaxy in the aperture but only an empty piece of sky. We would use UKT14 to measure the strength of the sub-mm radiation at both positions. Since in one case the measurement would include the radiation from the galaxy and the atmosphere and in the other case the measurement would only include the radiation from the atmosphere, the difference between the two would tell us the strength of the radiation coming from the galaxy alone.[****] The importance of the chart recorder, according to Gareth, was that we would be able to follow this whole process on the strip of graph paper, with the pen arm starting a new parallel red line each time the telescope moved. He said we would be able to use the lines to monitor the atmospheric conditions, with a flickering line showing the atmosphere was unstable and the data should not be used. He said we should note these periods by writing on the moving paper.

The other device was something I had seen before, a printer like those back in the office. The printer, according to Bill, would print out the signal and the noise during the observation. This I did understand. Signal and noise have precise technical scientific meanings, but these are not too far from the everyday ones. If I am in a coffee shop chatting to someone, the words from that person are the signal; the hubbub around me is the noise. If the ratio of signal to noise is high, I can understand what the person is saying; if the signal-to-noise ratio is low I can't.

[§§§] That's the way we stored electronic data in the 80s.

[¶¶¶] The atmosphere is completely opaque in the far-infrared waveband, but there are atmospheric windows in the near-infrared waveband (Figure 1.1). These were the ones used by Gareth and Eric.

[****] I have simplified this a little because there is a second motion that makes it possible to avoid the results being affected by changes in the emission from the atmosphere.

Noise can't be avoided in astronomy. There is no astronomical instrument anywhere that is not ultimately limited by noise, which, in modern cameras and sub-mm instruments like UKT14, consists of tiny fluctuations in the electrical currents and voltages in the instrument's electronics.

As in the coffee shop, the crucial thing is the signal-to-noise ratio, but there is one subtle but very important difference. In the coffee shop, I know whether I have detected the signal from whether I can understand the person. But with an astronomical instrument, you can never be certain you have detected a signal because both the signal and the noise manifest themselves as currents and voltages within the instrument. If the signal from some galaxy is higher than the fluctuations expected from the noise you have probably detected the galaxy – but only probably because what you hope is the signal might be a rare large noise fluctuation. There are a couple of values for signal-to-noise ratio used by scientists everywhere as useful rules-of-thumb. There are precise mathematical meanings but in everyday terms, a signal-to-noise ratio of 3 means that you have probably detected the galaxy and a signal-to-noise ratio of 5 means that you have almost certainly detected it.[††††] However, you can never be absolutely 100% certain.

Bill said that if all went well the signal-to-noise ratio printed out by the printer would gradually climb during an observation as more sub-mm radiation from the galaxy passed through the instrument's aperture. We would not need magnetic tapes to store our results. They would all be stored on sheets of printer paper. Even for the 80s, this did seem a bit old-fashioned.

After talking to Bill, we went down the mountain and went to bed. The following afternoon, Gareth and I sat at dinner planning the observations we were going to carry out that night. By this stage in my postdoc, Hale Pohaku was one of my favourite places. I always enjoyed sitting there chatting to the astronomers who were using the other telescopes. The food, served by friendly local guys, was hardly haute cuisine but it was good enough, and the view was sublime. We were surrounded on three sides by picture windows, with the view on one side being dominated by the vast bulk of Mauna Loa, an active volcano only 38 metres lower than Mauna Kea, with half its surface streaked with black lava flows. In the valley between the two mountains, which was often filled with clouds, I would occasionally see a truck inching its way along the saddle road from Hilo on one side of the island to Kona on the other, but on the slopes of Mauna Loa I never saw anything human at all. The mountain did not look like any mountain I knew; its gentle rise and dark lava flows reminded me of some science-fiction book I had read once, a mountain on a much bigger, older planet (Figure 2.7).

After dinner, the exodus began. Everyone headed for the trucks. As Gareth drove us up the rough track towards the summit I could see the dust rising from trucks already further up the mountain.

This time, we went straight into the control room. While we waited for the TO to start up the telescope, Gareth tried to convince me again about the beauty of the chart recorder.

[††††] If the signal-to-noise ratio is 3, the chance of the signal being a noise fluctuation is about 0.1%. If the signal-to-noise ratio is 5, the chance is only 0.00003%, and it is almost certain (but not completely certain) that the signal is real.

FIGURE 2.7   Mauna Kea Observatory. Mauna Loa is the mountain in the distance. Hale Pohaku is on the far side of Mauna Kea and the saddle road is between the two mountains. The United Kingdom Infrared Telescope is the one with the shiny silver dome in the foreground.

Credit: Institute for Astronomy/Mauna Kea Observatory

Although it was still not fully dark outside, the TO suggested that we check out UKT14 by observing Saturn. The telescope also had a standard optical camera, and in a few minutes, after some whirring of machinery from the dome, Saturn and its rings appeared on a TV screen. It was also now within the aperture of UKT14.

The TO started an observation. The printer started its typewriter rattle and, in a few minutes, we had detected Saturn with a signal-to-noise ratio of over 100. Planets are bright sub-mm sources.

The TO asked me for our first target. I found a galaxy on my list that was now almost overhead, the galaxy NGC 1614.‡‡‡‡ The machinery whirred. This time there was nothing on the TV screen – the galaxy was too faint. The TO started the observation with UKT14. The printer rattled to life.

I kept checking the numbers on the printer, but the signal-to-noise ratio never went above 1. After 20 minutes we gave up. I gave the TO the position of our next target, the galaxy NGC 3079, which is the galaxy whose far-infrared measurements are shown in Figure 2.6.

This time, after a few minutes there was the sign that we had something. The signal-to-noise ratio reached 2. It kept on increasing. It occasionally went down slightly but the overall trend was up. When the ratio reached 6.8 we decided to stop. We had made

---

‡‡‡‡ NGC stands for New General Catalogue, which is one of the standard galaxy catalogues.

the first observation at a wavelength of 350 micrometres. We now started a new one of the same galaxy at 450 micrometres. The signal was weaker, but the signal-to-noise ratio reached 4.3, still a solid detection. We went to an even longer wavelength, 850 micrometres, and this time the signal-to-noise ratio only reached 2.2, not even a 'probable' detection. We moved to a new galaxy.

The night continued. Gareth fell asleep. I gave up the pretence of using the chart recorder. I sat on top of the world huddled over the printer watching the results coming in. By the end of the night we had detected three galaxies, each at several different wavelengths.

On the following night, we detected four more. It was already a successful observing run and we had two more nights. But the gods of astronomy never give open-handed. Nights three and four were cloudy and we spent the nights sitting on the summit of a 14,000-foot quiescent volcano, reading, listening to music, talking, sleeping – a little boring but still romantic.

I was still on a high when I got back to the Institute. We had made some of the first-ever measurements of the sub-mm radiation from galaxies.

Then I discovered something worrying. I heard that a group in Germany had recently made some sub-mm observations of a similar sample of galaxies. The worrying thing was that their measurements of the strength of the galaxies' sub-mm radiation were roughly four times higher than ours.

During the next few weeks, I spent a lot of time in the library hunting for clues about which of us was right. Eventually, Eric suggested a possible explanation. He thought the problem might be the filters. The filters in UKT14 were made by the sub-mm gang at Queen Mary College, who were already known for the quality of their filters (the maestros of sub-mm filters – Chapter 1).§§§§ Eric suspected there might be a leak in the Germans' filters, which would explain why their measurements were so high.

I still wasn't sure, but I had run out of ideas, and Eric's explanation was at least plausible. I combined our sub-mm measurements with the IRAS far-infrared measurements to estimate the wavelength at which the radiation from the dust in each galaxy was strongest, and from that the temperature of the dust. I came up with temperatures between 30 and 50 kelvins, much higher than the 20 kelvins predicted by the theorists, which would have made an exciting conclusion for a paper.

Except by now I had realised another problem. Warm dust emits stronger sub-mm radiation than cold dust. We couldn't rule out the possibility that most of the dust in our galaxies was at 20 kelvins – exactly as the theorists had predicted – but that its radiation was being swamped by the radiation from a smaller amount of warm dust. Nevertheless, we had made some interesting observations. I wrote up our results as a paper for the Astrophysical Journal,[6] feeling quite pleased about my first trip across the sub-mm frontier.

For the next few years, I was quite excited about this new frontier. The telescope we had used was not a sub-mm telescope but one that had been designed to do infrared observations, so we had been borrowing somebody else's telescope like the sub-mm gang before us. But only a year later sub-mm astronomers finally got their own telescope. The James Clerk Maxwell Telescope, the first true sub-mm telescope, started operations on Mauna Kea in

---

§§§§ The Queen Mary group later moved to Cardiff University.

1987, and sub-mm astronomers no longer had to get out of the way of the real astronomers. It felt like the beginning of a new era.

I was the first scheduled observer on the new telescope and I went on several observing runs. But after a couple of years, I began to lose interest. The problem was that sub-mm astronomy really was primitive. Taking pictures is a fundamental part of astronomy – astronomers have been doing it in the visible waveband for 150 years – but the new telescope did not have its own camera. All it had was the same instrument Gareth and I had used, which had been moved over to the new telescope. With UKT14 it was only possible to measure the strength of the sub-mm radiation in a single direction. There was also the practical matter that I needed some results to get my next postdoc contract. I drifted off into other wavebands. It didn't seem likely that it would be possible to do real astronomy in the sub-mm waveband any time soon.

CHAPTER 3

Dust Stories

W HY DOES ANYONE BECOME an astronomer? I know why I did, or at least I have one or two answers I always give to people when I get asked this question, but I didn't know about anyone else until the interviews I did before writing this book.

Most of the questions I asked in the interviews were pretty impersonal, about *Herschel*, about the role the interviewee had played in the mission, about what they thought about the fate of the observatory. But I always ended with a few off-the-wall personal questions, questions that didn't seem strictly relevant to *Herschel* but were just ones to which I wanted to know the answers. Astronomy is a fairly social activity. We spend a lot of our time sitting alone in a room staring at a screen, but we also spend a lot of time talking to other astronomers. Most of the talk though is about research, jobs, grant applications, gossip about other astronomers, all pretty impersonal stuff – there is not much baring of souls. This was a chance, after softening them up with a lot of professional questions, to see what made a group of astronomers tick.

So, I do know the answer or at least the answers a group of astronomers (a cluster, a constellation?) give when they look back and try to make sense of their career.

I had two big surprises. The first of these was that everyone seemed to have been bitten by the bug at about the same time. When I asked the question about when they first became interested in science it was always around the age of five or six. The other big surprise was that it was never anything to do with formal education. It was always something that happened outside school. It might have been a parent who took them outside to look at the night sky. It might have been a book or a picture of the solar system on their bedroom wall. One scientist, who grew up to build key parts for one of the *Herschel* cameras, remembers helping her father, who loved making gadgets, build his own cameras and develop photographs in their garden shed. Another astronomer, with more mathematical tastes, remembers lying on the grass looking at the stars and thinking that numbers are like the universe because they go on forever. For many of us, it was science fiction. However hokey the story and characters, seeing a movie or TV programme in which humans are set among the stars and planets seems for many of us to have been the thing that set us on the

DOI: 10.1201/9781003195290-3

path to becoming an astronomer (perhaps it made the universe seem more real). Star Trek is probably responsible for more astronomers than anything else.

As for me, what would I answer? Sometimes I tell the story of hiding behind the sofa aged four, peeping out to watch the Daleks on TV. I sometimes mention the effect of the space programme; I was eight when I saw the pictures of the Earth rising over the Moon taken from Apollo 8, and it was only a year later that I saw humans walk on another world for the first time. But whether all this turned me into an astronomer, I am not sure. I was interested in lots of things when I was a kid, archaeology for one, and an honest answer would probably be that I am not sure what was the thing that got me started.

One answer I don't usually give because it is complicated is something that did happen at school. I remember when I was about ten being given a workbook on long multiplication, which included a question about the distance to the star Alpha Centauri, which is four light-years away. The question was to convert this distance into kilometres. The definition of a light-year is the distance light travels in a year, and the speed of light is 300,000 kilometres per second. So, in one minute light travels $60 \times 300,000$ kilometres, in one hour it travels $60 \times 60 \times 300,000$ kilometres, in one day it travels $24 \times 60 \times 60 \times 300,000$ kilometres, in one year it travels $365 \times 24 \times 60 \times 60 \times 300,000$ kilometres – and then multiply everything by 4 to get the answer. Doing this on a calculator is very easy, but this was before the days of calculators, and I remember it took a while. Doing it now on the calculator, I find the answer is 37,843,200,000,000 kilometres. I remember this had a big impact on me, but I don't remember whether it was because the number was so large or because of the idea that the distances of the stars are so large that we are looking back in time (If Alpha Centauri suddenly explodes we would not find out for four years.).

Astronomy seems for most of us to have been a vocation, but it should be admitted that it is also a solid middle-class career. There is some insecurity – very few people who start out doing astronomy PhDs end up as professional astronomers[*] – but it is nothing like the insecurity of some jobs. When we have a job we are paid well, and there are usually well-paid science and technology jobs waiting outside for those who have to leave the community.

Back in the nineteenth century, though, astronomy had not yet become a career, and nobody went into astronomy to make money. Of all the astronomers I have ever read about, the one who seems to have had the purest vocation, free of any interest in money, social prestige or any of the non-science perks of an astronomy career, was Edward Emerson Barnard. He was also the person who discovered the stuff I have spent my life observing.

Barnard's childhood sounds like it was taken from one of the novels of Dickens.[1] He was born in 1857 in Nashville, Tennessee. His father died before he was born and he spent the first ten years of his life in poverty and near starvation. His childhood overlapped with

---

[*] Even today very few people get paid to spend 100% of their time doing research in astronomy. But a lecturer or professor in a university probably gets to spend about half their time doing research, which is still a pretty good deal.

the American Civil War and for a time Nashville was blockaded. He later told the story of watching a steamer try to carry supplies along the river into the city. When the steamer was sunk, the onlookers dived into the river to rescue the supplies, and Barnard, who must only have been about five, managed to salvage part of a box of crackers, which his mother later whipped into batter and made into cakes. She scraped a living making wax flowers and was so poor she could not afford to send him to school. He claimed that in his whole life he only attended school for a total of two months. When he reached the age of nine, his mother sent him out to work fulltime at a photographer's studio. Barnard described his early youth as 'so sad and bitter that even now I cannot look back to it without a shudder'.[2]

When he was about 12 his mother's health or mind began to fail. From then on, it was his wages at the photographer's studio that supported them both. He spent much of his teenage years living alone in a boarding house. In a letter to one of the few people who was even aware of him at this time, he wrote: 'Clinging to me through life has ever been the memory of [your] kind word and nod and smile of recognition to a poor sick ragged boy on his way to or from work....This is not sentiment. It is plain and substantial reality'.[3]

From these grim beginnings, he somehow became an astronomer. At the age of 19, he happened to read a book on astronomy that a friend had left with him as security for a loan. With the money he was earning at the photographer's studio, he bought a telescope, and with it he began to discover comets. A philanthropist, H.H. Warner, had offered a prize of 200 dollars for every comet discovered by an American, and Barnard picked up several of these prizes, which he used to buy a house for himself, his invalid mother, and his new wife. Eventually, he was offered a fellowship at Vanderbilt University, which was less money than his job in the photographer's studio but allowed him to take up astronomy full time. His fame as an observer spread and he was offered a job at Lick Observatory on Mount Hamilton in California.

Lick was the first of the new mountaintop observatories, from which the view of the sky is much crisper than from sea level. At the time, Lick also had the biggest telescope in the world, the 36-inch telescope.[†] So the job was a huge opportunity. Unfortunately, Barnard's pure vocation seems to have been associated with a complete inability to compromise and he soon fell out with the director of the observatory, Edward Holden. At an observatory so isolated that all supplies were brought in by mule, the relationship between the two became so bad that after five years they were communicating only by written notes. Eventually Barnard decided to leave, but while there he made some big discoveries.

At the time he was regarded as one of the greatest observers who ever lived.[4] His eyesight seems to have been phenomenal, which was crucial in an era when astronomers still looked through telescopes. Most people can see seven stars in the Pleiades star cluster[‡] – Barnard claimed he could see 12. He also had immense physical endurance. At Yerkes Observatory near Chicago, in his later career, he spent nights in the telescope dome when the temperature dipped down to −26 degrees Centigrade.[5] His drive was legendary. He wanted to spend every second possible looking through a telescope and he was known at Lick as 'the

---

[†]　36 inches was the diameter of the mirror.
[‡]　Another name for the Pleiades is the 'Seven Sisters'.

man who never sleeps'. One of the reasons he despised Holden was that the director kept the 36-inch telescope for himself, but even when the sky was clear he would often still go to bed at midnight, which even I as a pampered modern astronomer think is bad form. During Barnard's first three years at Lick he also didn't take a single holiday. (Nobody ever asked his wife, who was stuck with him on the mountain, a mule-ride away from civilisation, what she thought about any of this.)

When asked about how he became interested in astronomy, he said that he remembered when he was four seeing the great comet of 1861. He also said that he used to lie in the back of a wagon watching the stars which, he claimed, 'helped to soften the sadness of my childhood'.[6] His answer then was not too different from the answers of my modern astronomers. Barnard was one of us.

During his years at Lick, Barnard measured the opacity of Saturn's rings and discovered the largest mountain in the solar system, Olympus Mons on Mars. He was also the first to show that a nova, the sudden appearance of an apparently new star, is the result of a stellar explosion. And, 400 years after Galileo had discovered the four large moons of Jupiter, he found a fifth moon, Amalthea, a discovery which made him a global celebrity. Some of these discoveries today do not seem quite so extraordinary. We know that Jupiter has at least 79 moons, so the discovery of a fifth one now seems small potatoes.[§] One of his discoveries, though, is still of huge importance today.

Barnard's early life may have been blighted by ill fortune, but in his professional life he had two big slices of luck. The first was that he started his career at just the moment that photography became important in astronomy, the second that his years in a photographer's studio had given him all the necessary technical knowledge.

When he arrived at Lick, he found that all the big telescopes were reserved for senior astronomers like Holden. But he realised that photography changed everything – with a camera even a small telescope could be useful. He decided to explore the Milky Way.

If there are no streetlights around, the Milky Way is easy enough to see, a faint band of light that stretches across the sky. The Milky Way is the Galaxy seen edge-on, the combined light from millions of stars that we see, from our position inside it, when we look along its disk. But if you forget for the moment that you know this, the Milky Way doesn't really look much like the cross-section of a disk. It is much more irregular than that. It actually looks more like what the Romans thought it was: a jug of milk splashed across the sky.

It was William Herschel who was the first to show that when viewed through a telescope the smooth light of the Milky Way dissolves into a myriad of stars. He was also the first person to try to map the structure of the Galaxy by counting the number of stars in different directions (Chapter 7). Something that surprised him is that there are some dark areas in the Milky Way where there are hardly any stars at all, which he called 'holes in the heavens'.[7] Herschel thought these dark areas showed the absence of stars. The aboriginal Australians thought these dark areas, which are more obvious from the southern

---

§   Appropriately, Amalthea does look rather like a small potato.

FIGURE 3.1   The Australian Aboriginal Emu in the Sky above a rock engraving of the Emu in Ku-Ring-Gai Chase National Park.

Credit: Barnaby and Ray Norris

hemisphere, showed the presence of something, imagining that one particularly spectacular dark shape was a giant emu sprawled across the sky (Figure 3.1).[8]

Barnard realised photography was a game changer. Even with a small telescope, it would be possible with a camera to observe much fainter parts of the Milky Way than is possible with the naked eye. For observing faint diffuse structure like the Milky Way the crucial property of a telescope is its field-of-view,⁵ the area of sky visible through the telescope. The big telescopes at Lick only had a small field-of-view, which was fine with Barnard because

---

⁵   In technical terms, the crucial property is the telescope's f-number: the ratio of the diameter of its lens or mirror to its focal length. The lower the f-number, the larger the telescope's field-of-view and the better it is for seeing faint diffuse emission. A small telescope is equally as good for observing faint diffuse emission as a large telescope with the same f-number.

FIGURE 3.2   Barnard's homemade telescope (the box with the tube sticking out of it) attached to the 6.5-inch telescope. Barnard (shown in the picture) kept his homemade telescope from drifting out of position by observing a star through the 6.5-inch telescope, nudging the telescope back into position if the star drifted away from the centre of the field.

Credit: Special Collections, University Library, University of California Santa Cruz. Lick Observatory Records

he couldn't get any observing time on them anyway. But the small telescopes at the observatory also had a small field-of-view. He decided to make his own.

He knew from his time in the photographer's studio that the lenses used in portrait photography (all those formal pictures of grim Victorians!) have a large field-of-view. He made a simple telescope out of one of these portrait lenses, using a photographic plate, a glass plate covered in photographic emulsion, to record the picture – all enclosed in a simple wooden box. To steer this homemade telescope around the sky, he mounted it on one of the regular telescopes at the observatory, the 6.5-inch telescope (Figure 3.2).[9]

By modern standards, Barnard's exposure times were crazily long. Nowadays it is rare to use an exposure time of more than an hour, and for a modern astronomer a long exposure is an opportunity to wander away from the telescope for a bit – have a coffee, read a book.

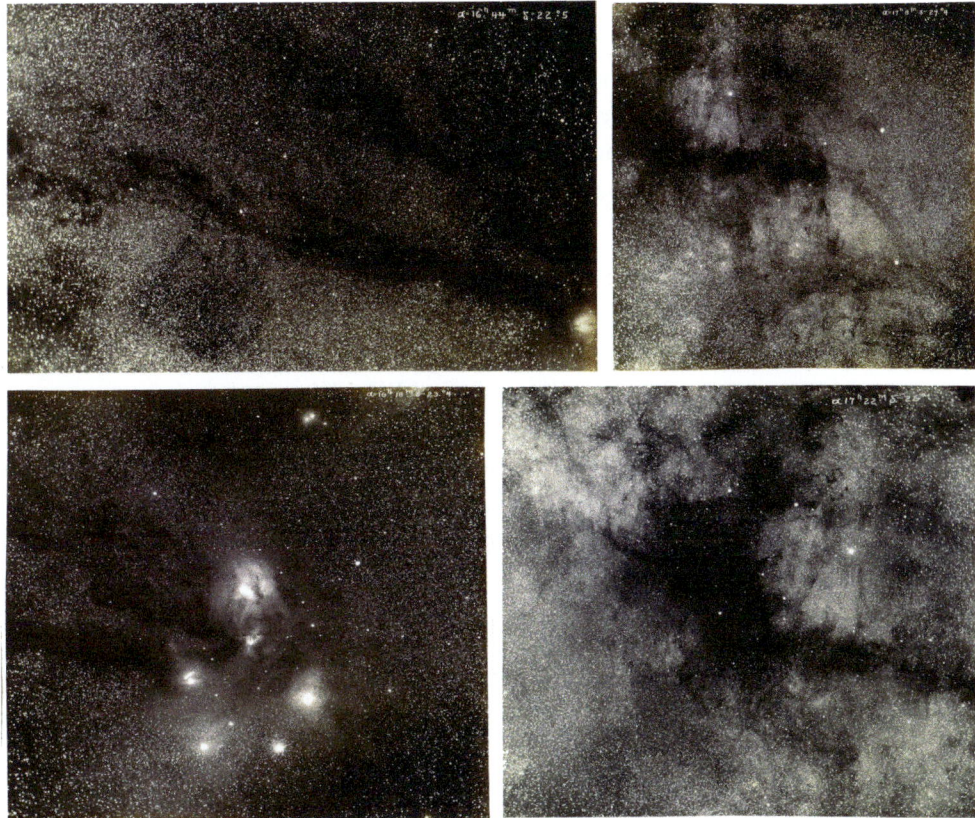

FIGURE 3.3   Four of Barnard's photographic plates of the Milky Way.

Credit: Carnegie Institute for Science

Because the Milky Way is so faint, and a photographic plate was much less sensitive than the CCDs in today's cameras, Barnard had to use much longer exposures. He also had to be there at the telescope the whole time. Left to itself, a telescope will slowly drift out of position, and in the nineteenth century the standard way to avoid this was for the astronomer to use a second telescope, a guide telescope, to observe a bright star, nudging the telescope back into position if the star drifted from the centre of the field. Barnard used the 6.5-inch telescope itself as his guide telescope, taking his pictures with his homemade telescope mounted on its side (Figure 3.2).

On the night of 1 August 1889, he took his first successful picture of the Milky Way with an exposure time of three hours and seven minutes. All that time, he had to stand there, staring at a star through the 6.5-inch telescope, alone in the dome. And this was one of his shorter exposures – some were almost six hours long. The pictures he took revealed for the first time the structure of the interstellar medium (Figure 3.3).[**]

[**]   All his beautiful images can now be found on the web at https://exhibit-archive.library.gatech.edu/barnard/project_info.html.

Barnard must have realised immediately the importance of the dark structures he saw in his pictures because he persevered with this observing programme, which must have been physically exhausting. He also spent a lot of time and money trying to get the pictures published, a major challenge in the early days of astrophotography. He wrote extensively about these pictures, both in science journals and for the public, which also suggests he knew their importance. Here is what he wrote about his very first picture:

> This remarkable picture shows the cloud-forms like waves of spray. A dark curving lane runs from the lower left-hand portion of the picture and curves gracefully upwards to the place of Jupiter. It is singularly like the stem of a great leaf. At the middle of the picture it is seen to pass behind some of the clouds of stars and emerge beyond, showing us clearly which part of the Milky Way at that point is nearest to us. Imagination may aid one, but it looks as if the lines of the cloud-forms, and of the stars and vacancies, all run more or less concentric with this extensive lane. In the reduced glass positive, this lane and its connected branches come out grandly dark…[and] the mass of stars to the north-west of the middle resemble breaking spray.[10]

In all his writing about his pictures of the Milky Way, he used terms from the natural world – spray, stem, leaf, cloud, branch – which suggests he thought he had discovered something physical out there in interstellar space. The thing I find puzzling is how long it took him to accept the truth of this. For almost 20 years, he maintained, like William Herschel, that the dark shapes in his pictures were simply places where there were no stars.

The person who realised their true explanation is someone whose name is almost completely forgotten now. Arthur Cowper Ranyard came from a comfortable middle-class background. He did a degree in mathematics at University College London and then trained and worked as a lawyer. His real passion, though, was astronomy. Living alone in a house in London, he spent all his free time on his scientific interests. In 1888 he became editor of the popular science magazine *Knowledge*, a magazine that had been founded to cater to the public hunger in late-Victorian England (think gaslight, hackney cabs and Sherlock Holmes) for information about the latest scientific discoveries.

Even before he knew about Barnard's pictures, Ranyard was sure there was something physical in interstellar space. In 1889 he visited Burlington House in Piccadilly, the home of the Royal Astronomical Society, to see a set of drawings of the Milky Way made by the German astronomer Otto Boeddicker. Boeddicker had not used a camera but had observed the Milky Way through an opera-glass[††] and made drawings of his observations.

Ranyard was fascinated by the drawings. He wrote for his readers in *Knowledge* that they consisted of 'whisps and streams of light with very numerous dark channels, having more or less sharply defined edges'. The dark channels did not look to Ranyard like places where there were no stars but like 'opaque matter, dust clouds or fog-filled space, which cut out the light of the bright streams [of stars beyond]'.[11] (Reading through the issues of

---

[††] An opera glass is designed to get a close-up view of an opera so has a large field-of-view.

Knowledge[‡‡] today is like being transported back into a Victorian club. In the same issue as Ranyard's article there were also articles with titles: The Common Cockroach, The Fish Lizards of Secondary Rocks, The Ethnological Significance of the Beech, Some Properties of Numbers, Colour Blindness, On Large Telescopes, and Barnacles.)

When Ranyard heard about Barnard's pictures, he began to publish them in *Knowledge*, with the first one appearing in the issue for 1st July 1890. For the next four years, Barnard and Ranyard carried on a lively debate in the letters section about the explanation of the dark structures in Barnard's pictures. They reinforced Ranyard's belief there must be something in interstellar space obscuring starlight. In the issue in which he showed Barnard's first picture, he wrote:

> That some of the light from this direction of space is absorbed by dark masses, or regions rich in meteoritic dust, is rendered probable by many considerations…. The many small isolated dark patches on a comparatively uniform background light are much more easily accounted for as produced by the absorption of foggy areas, or dark bodies, than as holes or gaps in the stellar clouds.

Barnard didn't agree. In the issue for 1st January 1894, he wrote:

> Looking at these peculiar features, I cannot well see how one can avoid the conclusion that they are necessarily real vacancies in the Milky Way, through which we look out into the blackness of space. I am aware that you are opposed to this view, and I would like to have your view of the real nature of these apparent crevices in the Milky Way, as shown on this particular plate.

In an issue that appeared later the same year Ranyard responded with a subtle argument, which I think should have been conclusive. Suppose that, as Barnard claimed, there are tunnels running from one side of the Galaxy to the other. From our perspective inside the Galaxy, we would only see the multitude of dark areas that we do if the solar system happened to lie at the intersection of a very large number of tunnels.

It seems unlikely that so many tunnels would intersect in one place, and it seems doubly unlikely that the solar system would happen to be at that place. The debate came to an end later that year when Ranyard died at the early age of 49. Barnard was still not convinced.

It was only in 1913, 19 years later, that Barnard finally accepted that Ranyard must be correct. He was observing the Milky Way from Yerkes Observatory one beautiful summer night:

> I was struck by the presence of a group of tiny cumulous clouds scattered over the rich star clouds of Sagittarius. They were remarkable for their smallness and definite outlines – some not being larger than the moon. Against the bright background they appeared as conspicuous and black as drops of ink. They were in every way like the black spots shown on photographs of the Milky Way, some of which I was at that moment photographing. The phenomenon was impressive and full of

---

[‡‡] There is an archive of all the issues of *Knowledge* at the HathiTrust Digital Library (https://babel.hathitrust.org/cgi/pt?id=uc1.c2834662&seq=7).

suggestion. One could not resist the impression that many of the small spots in the Milky Way are due to a cause similar to that of the small black clouds mentioned above – that is, to more or less opaque masses between us and the Milky Way.[12]

Why did he take so long to accept what now seems obvious? His enemy Edward Holden said that Barnard had the same gift as William Herschel, the other great observer of the nineteenth century, for *'seeing* and noting what is new'. But he added that 'his reasoning on what he sees is very apt to be quite wrong'.[13] Of course, it was Herschel who had first claimed the dark areas were places where there were no stars, 'holes in the heavens', and possibly it was just too difficult for someone with only two months of education to accept that the great Sir William Herschel had been wrong.

By this time, many astronomers had accepted there must be something in interstellar space obscuring starlight. Back in the 1890s, both Ranyard and Barnard had used the term 'dust', possibly inspired by the explosion in 1883 of Krakatoa, which threw volcanic dust into the atmosphere, dimming the Sun and producing spectacular red skies.[14] In 1922, in a beautiful little paper less than two pages in length,[15] the American astronomer Henry Norris Russell showed that if there were tiny solid particles in interstellar space, they would efficiently absorb and scatter starlight. Acting like cosmic smoke, they would make the stars appear both dimmer and redder. The final touch was placed in 1930 by the Swiss-American astronomer Robert Trumpler.[16] Trumpler showed that distant star clusters were both fainter and redder than expected, which can be explained if space is filled with tiny solid particles – interstellar dust grains.

Who discovered interstellar dust? Most astronomers claim it was Trumpler, and it is true that he found the first quantitative evidence for its existence. Four decades before this, Barnard obtained pictures that showed its existence – if one reasons correctly, which he didn't. I like to think that it was Ranyard. He died relatively young, and his name is now almost completely forgotten, but he realised, even before he saw Barnard's pictures, that there must be something in space obscuring starlight. He even used the word *dust*. But perhaps I am just being a romantic.

Dust is everywhere. Every galaxy contains dust, and the dust in galaxies is the source of much of their beauty. The dark band around the middle of the Sombrero Galaxy (Figure 3.4) is caused by dust, which is hiding the light from the stars behind. It is the dust that is responsible for the network of veins in the spiral galaxy Messier 83 (Figure 3.5). There is even now evidence that dust is found outside galaxies.[17] And closer to home, dust grains are responsible for some of the prettiest pictures of objects in our own galaxy.

Figure 3.6 shows the prettiest picture of all – perhaps – the Pillars of Creation. The picture is a *Hubble* image, but the true artist is the dust. The dust in the pillars is absorbing the light from the glowing gas behind, leaving the dust etched against the colourful backdrop. The beauty is ephemeral because radiation from hot stars, just off the edge of the image, is gradually eroding the pillars. Eventually they will vanish. These really are pillars of creation because deep inside them, hidden by the dust, there are stars being born.

FIGURE 3.4   Image of the Sombrero Galaxy made with the *Hubble Space Telescope*. The dust in the galaxy's disk is hiding the light from the stars behind.

Credit: NASA and the *Hubble* Heritage Team (STScI/AURA)

FIGURE 3.5   *Hubble* image of the spiral galaxy Messier 83.

Credit: NASA, ESA and Z. Levay (STScI/AURA)

FIGURE 3.6   The prettiest picture in astronomy? The *Hubble* image of the Pillars of Creation.

Credit: NASA, ESA

For the true dust nerd, the big problem about living on a planet is that there isn't much interstellar dust around. The planets and all the other objects in the solar system were formed 4.5 billion years ago out of a hot disk of gas and interstellar dust, but most of the dust that was originally in the disk would have been volatilised by the heat.

For a while, astronomers hoped that some interstellar dust might have survived in the comets, which were formed further out in the disk where the temperature was lower. In 1999, NASA sent a spacecraft out to have a look. In 2004, *Stardust* arrived at Comet Wild 2, scooping up some of the solid particles that had been ejected from the comet and taking them back to the Earth. Back on Earth, though, when the scientists analysed them, they found that most of the particles had condensed out of the gas as the disk cooled[18] and had not been there before the formation of the disk. They were not true interstellar dust grains.

Luckily there were a few. The scientists who do these things, using techniques that are well outside my skillset as an observational astronomer, used the presence of certain radioactive isotopes to show that a few of the particles did seem to be genuine interstellar dust grains.[19] The same is true of meteorites. Within a meteorite, there are often a few tiny particles that contain radioactive isotopes that imply they were there before the beginning of

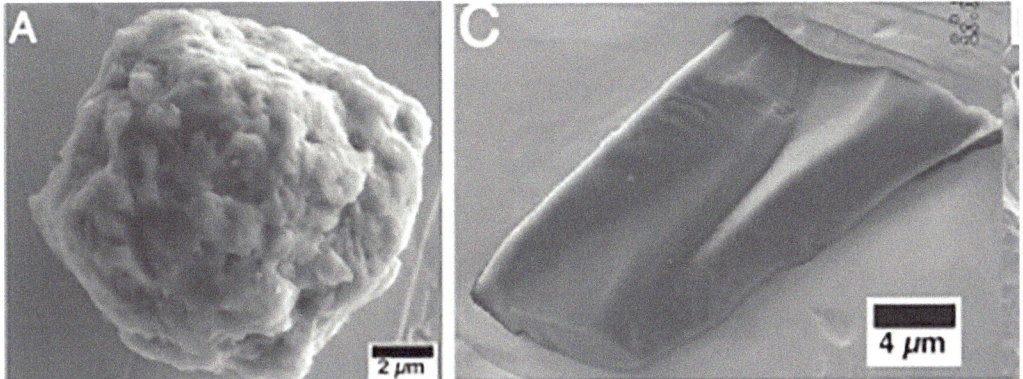

FIGURE 3.7   Two interstellar dust grains found in a meteorite that fell at Murchison Australia in 1969. The one on the right is probably a shard broken off the original grain. The two rulers in the bottom right of each panel show lengths in micrometres, one thousandth of a millimetre.

Credit: Phillip Heck

the solar system – interstellar dust grains that were swept up in the disk and managed to survive the intense heat. In the solar system, there is not a lot of interstellar dust around, but there is some.

In an investigation which I find almost unbelievably impressive, scientists have managed to reconstruct the life stories of two interstellar dust grains. Figure 3.7 shows these grains, which were found in a meteorite that fell in 1969 close to Murchison, Australia. Using a set of forensic techniques I don't claim to understand, a team led by Phillip Heck of the Field Museum of Natural History in Chicago has deduced their history.[20]

About 4.7 billion years ago, there was a star close to the end of its life with a mass about twice that of the Sun. When it ran out of nuclear fuel in its core, its outer layers swelled up and cooled and it became a red giant star. As its outer layers cooled, solid particles – dust grains – condensed out of the gas. The pressure of the star's radiation propelled the two dust grains out into space.

For the next 100 million years, not much happened to the them. Eventually, while they were passing through a dense cloud of gas and dust, the cloud collapsed, forming a disk. As the disk cooled, objects coalesced out of the disk. The two dust grains might have ended up in a planet or comet, but they happened to end up in an asteroid.

For another 4.5 billion years, not much happened. The asteroid then collided with another asteroid, knocking off a fragment which happened to include the grains. The fragment fell to earth as a meteorite and eventually the grains ended up in a laboratory in Chicago – the end of a rather splendid journey. The reason they were able to survive the heat of the disk is that they are composed of carborundum, silicon carbide, one of the toughest substances known.

Most interstellar dust grains, though, did not survive the heat. We cannot therefore find out the composition of interstellar dust by simply analysing the grains in meteorites because these are only the survivors. The reason we do have some idea of the composition of the dust that was lost is because of one of the most useful techniques in astronomy.

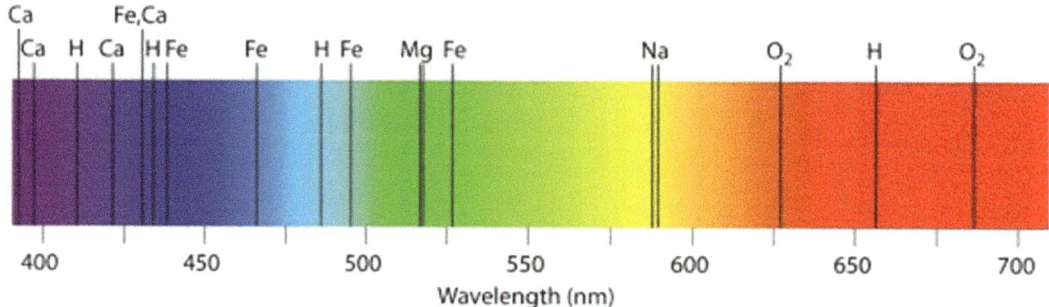

FIGURE 3.8   A spectrum of the Sun. Wavelength, which is plotted along the bottom, is measured here in nanometres, a nanometre being one billionth of a metre.

The light or any other kind of electromagnetic radiation from an object contains a blend of wavelengths. Spectroscopy is the technique of unblending the wavelengths. There are various ways you can do this. The simplest method is to use a glass prism, as William Hershel did (Chapter 2), which, by bending the light of different wavelengths by different amounts, unrolls the light into a multi-coloured carpet – a spectrum. Astronomers nowadays have other devices for doing this,[§§] but they all work in the same way by bending light (or other types of electromagnetic radiation) by an amount that depends on its wavelength.

In the early nineteenth century, a German physicist, Joseph von Fraunhofer, was studying the spectrum of sunlight, doing much the same thing as Herschel although with more sophisticated equipment, when he noticed the spectrum was crossed by a series of parallel dark lines (Figure 3.8). It took almost a century for scientists to figure out the explanation, but eventually it was realised that these lines are a consequence of the strange quantum world inhabited by atoms, electrons and other subatomic particles. One of the laws of this inapprehensible world – describable by equations but not accessible intuitively to us in our world one million billion times larger – is that atoms are only allowed to jump up or down in energy, by absorbing or emitting a photon of radiation, by strictly defined amounts. Another of the strange laws of this world is that there is a mathematical relationship between the energy carried by a photon of radiation and its wavelength, so a corollary of the first law is that an atom is only allowed to emit or absorb radiation at certain specific wavelengths. The dark lines in the Sun's spectrum show the wavelengths at which the atoms in the Sun are allowed to absorb light.

The first reason that spectroscopy is such a fundamental part of astronomy is that it is our only way of finding out what the objects out there in space are made of, apart from going there which isn't practical outside the solar system. There are a different set of allowed wavelengths for each chemical element, which effectively act as the fingerprints of the element. The dark lines in the Sun's spectrum – the spectral lines – are the fingerprints that tell us which elements are present there. The spectrum in Figure 3.8 is only part of the full solar spectrum, but even this tiny part shows that the Sun contains hydrogen (H), iron (Fe),

---

§§   Astronomers today often use a grating, a series of parallel lines etched on a glass sheet, but it does exactly the same thing as Herschel's triangular slab of glass.

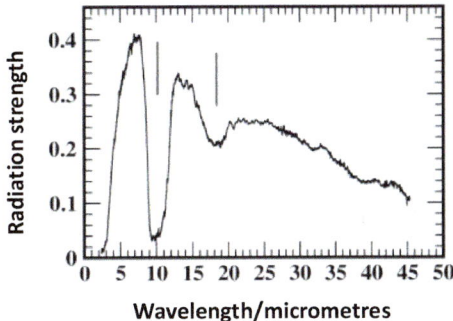

FIGURE 3.9   A spectrum of interstellar dust. The two dips in the spectrum, indicated by the vertical lines, are the spectral lines from chemical bonds in the minerals that make up the dust grains.

magnesium (Mg), sodium (Na) and oxygen molecules (O_2). Occasionally there are the fingerprints of someone we did not know about - helium, the second most common element in the universe, was only discovered because of its spectral lines in the Sun's spectrum.

Spectroscopy is also fundamental because it allows us to measure the speed of things. It has been known for almost 200 years that the wavelength of radiation changes if the source of the radiation is moving, with the size of the change being proportional to the speed of the source.[**] An everyday example is the sudden change in the pitch of an ambulance's siren as it passes us on the street, shifting from moving towards us to moving away from us. We know the wavelengths of the elements' spectral lines from measurements on Earth, so by measuring the changes in their wavelengths when we observe them in some galaxy or star, we can estimate the object's speed – and that allows us to do all kinds of wonderful things, from measuring the mass of a galaxy to uncovering the history of the universe itself (Chapter 4).

Spectroscopy is the golden thread that leads us out into the universe. Without spectroscopy, we would not know any of this.

Spectroscopy is also why we know something about the composition of interstellar dust. Figure 3.9 shows a spectrum in the infrared waveband of some interstellar dust, although plotted as radiation strength versus wavelength rather than in the form of an image as I used to show the spectrum of the Sun. The dips in the spectrum at wavelengths of around 10 and 18 micrometres are also spectral lines, although, in this case, these are the broader spectral lines produced by the energy absorbed by the chemical bonds between atoms rather than by the atoms themselves. These two spectral lines show the presence in this dust of the bond between a silicon and oxygen atom. The main chemical compounds containing this bond are the silicates, which are the minerals that make up most of the Earth's crust and mantle, so it seems that this dust at least is made of the same stuff that makes up our planet. Spectroscopic observations of other dust show that some dust grains are made up of carbon-rich compounds, and spectroscopy of dust deep inside dense clouds of gas and dust shows that the dust grains there are often encased in ice.

**   Called the Doppler shift after the scientist who discovered it.

Spectroscopy may be the golden thread, but it can only take us so far. We have a rough idea of what interstellar dust is made of, but we don't know many of the details. We know that a lot of interstellar dust is made of silicates, but we don't know what kind of silicates, for example. The same is true of many of dust's other properties. We know there are probably grains with a diameter of about one micrometre because grains of this size have the maximum reddening effect on light, but it is quite possible that most of the dust is made of bigger grains – we just don't know. We don't know whether the dust grains are little spheres or whether they are spindles, disks or are maybe even fluffy like snowflakes. We don't know whether a dust grain is made from a single mineral or from many. We don't know how a dust grain is made, what happens to it as it travels through space, and how it is destroyed.

If only we could travel out into interstellar space and pick some up.

I have never gone to work feeling bored. Occasionally it might be good to go to work, clock off at five and not think about it again. It has never been like that for me. I generally start thinking about work the moment I wake up and sometimes lie awake at night thinking about some research problem. It has not always been so good for those close to me. My wife talks about the 'two-second delay' when she is trying to talk to me and I am thinking about work.

I have never regretted becoming a sub-mm astronomer. Being there at the beginning, watching as one revolutionary instrument after another takes us further across the frontier, discovery following discovery, has been much more fun than sticking in some long-established resesarch field where most of the big discoveries were made long ago. It has also been fun to be part of a gang, even if not a very cool one. There are probably ten times more astronomers who work in the traditional visible waveband than sub-mm astronomers, and most sub-mm astronomers are still found in a few unfashionable institutions. I still work with people who are really physicists and only do astronomy because it gives them an excuse for building cool instruments. We mostly don't work in the elite institutions – Cambridge, Oxford and Imperial – but in places like Cardiff, the University of Central Lancashire, the University of Hertfordshire.

It doesn't help that the only thing many people know about us is that we observe interstellar dust. Tiny particles in interstellar space do not seem one of astronomy's big themes. To many optical astronomers, dust is just the irritating thing that absorbs visible light, something whose effect we need to correct, but no more interesting than that.

But dust is much more important than its effect on a few *Hubble* images. Figure 3.10 shows the total radiation from everything in the universe plotted against wavelength. Everything means everything, not just the radiation from objects that are bright enough to detect as individual objects but also the radiation from objects too faint to detect individually but that still contribute a tiny amount to the combined radiation. Measuring the total radiation from everything is quite tricky, but it can be done using specially designed telescopes. And everything doesn't just mean the objects around us in the universe today. When we look out into space, we are, courtesy of the speed of light, also looking back in

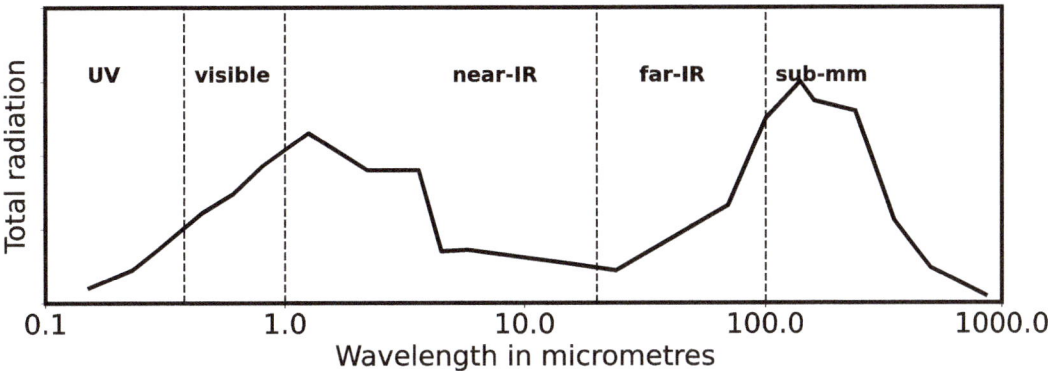

FIGURE 3.10   The total radiation from *everything* plotted against wavelength.[21]

time, from nearby stars like Alpha Centauri, which we see a few years back in time, to galaxies so far away that we are looking back almost to the big bang itself. Everything means the radiation from *everything*, from the objects around us today back to the first galaxies and stars.

There are two peaks in the figure. One, at the boundary of the visible and near-infrared wavebands, is the starlight, the combined light from all the stars in all the galaxies back to the big bang. The other, in the sub-mm waveband, is from interstellar dust, the combined radiation from all the dust in the universe back to the big bang. The two peaks are roughly the same size, which means that the combined radiation emitted by all the dust, ever, is roughly the same as the combined radiation from all the stars, ever. The radiation from the dust is also mostly starlight, but it is starlight that has been transformed by the dust: the dust grains absorb starlight; the energy in the starlight heats the dust grains; and the dust grains radiate the same amount of energy in the sub-mm waveband (like money-laundering – the amount of energy stays the same, but its form has changed, concealing the source). The equal size of the peaks therefore shows that roughly half of all the light ever emitted by stars, all the way back to the stars in the first galaxies, must have been absorbed by interstellar dust. The effect of dust on visible light is therefore more than the creation of a few beautiful *Hubble* images. The dust in the universe has absorbed half of all the light emitted by stars, ever.

Telescopes that work in the visible waveband like *Hubble* are therefore not enough because they can't see the missing half of the radiation that has been absorbed by the dust. Sometimes this doesn't matter too much, but for some of the biggest questions in astronomy it is a big deal. Stars are born in dense clouds of gas and dust and the dust hides their births almost completely from optical telescopes. Some galaxies also seem to be born in massive clouds of gas and dust (Chapter 11). It is therefore impossible to answer big questions like how the stars and galaxies were born without using sub-mm telescopes.

It still does feel, though, that optical astronomers don't take us seriously. I suspect the problem may be that optical astronomers think that what they observe with their telescopes is somehow more real because it is what we might see with our eyes. There isn't a word for this, but here's one I have just made up: visualism, short for visual chauvinism.

Visualism is probably worse now because of the beautiful *Hubble* pictures, in which objects do look like what you imagine you would see with your eyes if you could only get close enough. But the truth is, they are as false or true as the pictures in every other waveband.

It is not just the *Hubble* pictures. Early in the era of photography, astronomers stopped bothering to convert their photographic negatives, in which stars appeared as black against a white background, into photographic positives, in which stars appear white against a black background. Black or white, the information about the star is just the same, and it is often easier, for physiological reasons I don't understand, to see a faint object as a dark smudge against a light background than as a light smudge against a dark background.

In the digital age, everyone is an artist. In a camera, in the visible or any waveband, radiation photons are recorded as electrical signals. Astronomers are free to turn these signals into a picture in any way they like. Astronomers who work outside the visible waveband like me are the freest because we have no preconceptions of what our images should look like. I personally have quite conservative taste. I like to display a sub-mm image using a greyscale: dark grey where there is lots of sub-mm radiation, white where there is none. But astronomers use all kinds of funky palettes to represent the strength of the radiation. Optical astronomers are more inhibited because they usually have an idea in their mind of what the object should look like; they generally choose colours to make the images look like what they imagine the eye would see.

But take the stunning *Hubble* image of M83 in Figure 3.5. The human brain only sees colour if there is enough light to activate the cone cells in the eye's retina. There is no place in the universe where you could stand and receive enough light from M83 to see it in the colourful way it is portrayed in the *Hubble* image. The human eye sees galaxies in black and white.[22]

Visualism is therefore yet another anthropocentric fallacy. A sub-mm, or a radio or an X-ray image, is just as real as an optical one. The images provide different views of the universe, different perspectives.

Sub-mm astronomers stand tall, be proud of what you do! Here is an even better argument for why dust is important.

Stars form out of clouds of molecular gas, which are mostly made of hydrogen molecules, two atoms of hydrogen bound together. The only reason hydrogen molecules exist is because of interstellar dust. If there was no dust, hydrogen atoms would wander around interstellar space forever without finding each other. The hydrogen atoms stick to the dust grains and while they are there become attached to other hydrogen atoms, which is how molecular hydrogen is made. A molecular cloud would not exist without dust. The birth of a star requires a molecular cloud; planets are only formed around stars; and life, at least as we know it, seems to need planets.

We would not exist if it were not for interstellar dust.

CHAPTER 4

Whiteboard Memories

I THOUGHT ABOUT SUB-MM ASTRONOMY again when I got a permanent job. My contract in Hawaii was for three years. I got married to Keirsten shortly after moving and during our three years there we spent a lot of time discussing where we would go next. As my contract came towards its end, I applied for postdocs all over the world. Keirsten said what would we do if there were several job offers – how would we choose? I said, I didn't think that would be a problem.

And that's how it turned out. I was offered a single postdoc, at the Space Telescope Science Institute in Baltimore, which fortunately was one of the better places I had applied to. But then, just before we were due to fly to Baltimore, my application to renew my US visa was rejected. Immediately, everything fell to pieces. My salary in Hawaii was about to stop, we were paying expensive rent, we had just had a baby, we were thousands of miles away from any family, and now I had no job to go to. I applied for a new visa, but we made the contingency plan that if it didn't come through in time, we would use the last of our money to fly to Toronto and live in Keirsten's parents' basement. At the very last moment, as I was on the way to the airport to pay the balance of our Toronto tickets, I stopped off at the Institute. There was a message. The new visa had come through. We moved to Baltimore.

Even before *Hubble* had been launched, the Space Telescope Science Institute was a splendid place to work, but my brush with the US Immigration Service (the IRS without the people skills) had made us want to move out of the States. After a year, I got a temporary teaching position at the University of Toronto, filling in while someone went off to spend five years being a dean. From a personal point of view, the years in Canada were great. We lived in Keirsten's hometown and she was able to go to university. But from a professional point of view, things were slow. In my first two years in Toronto every observing trip except one was either wiped out by the weather or by instrument failure. The exception was when I had the chance to go observing on the Palomar 200-inch telescope. The 200-inch is an iconic telescope. It was the biggest telescope in the world for 40 years, and it was used by many of the greats in the history of astronomy. It was exhilarating to be able to ride up to the prime-focus cage in the same rickety elevator that Edwin Hubble himself must have used.

DOI: 10.1201/9781003195290-4

However, by the time I got there, the lights of San Diego had come too close; the night sky was too bright, and the data I got was useless.

The Toronto position was temporary, and I was still applying for jobs. But with no interesting research to show, I wasn't even making it onto shortlists. I began to think of a job outside astronomy. Unfortunately, it was pretty obvious that employers would not be lining up to give me one. Most of what I had been doing in the ten years since leaving university would look useless to any employer. The only realistic option was becoming a schoolteacher. And for that I would need training, which meant money, which we still didn't have.

Then things began to get better. I and a friend from Oxford, Steve Rawlings, got observing time on the United Kingdom Infrared Telescope, the same telescope that Gareth and I had used to observe the sub-mm radiation from some nearby galaxies (Chapter 2). The telescope now had a shiny new infrared spectrometer, and Steve and I used it to detect spectral lines from some distant radio galaxies, which at the time seemed quite exciting. We wrote a paper that was published in *Nature*, which is one of the two most prestigious journals for scientific papers.* I began to get on shortlists. Eventually, scarily close to the end of my Toronto contract, I got offered a permanent job as a lecturer in Cardiff. We moved back to the UK.

A permanent job changes everything. Until then you are chained to the postdoc cycle. You get to your new institution, if you are lucky with a three-year contract. Year 1 is spent settling in. Year 2 is the year when you are most productive. But by the end of Year 2, you need to start thinking about the next contract. Year 3 is mostly spent writing job applications. A postdoc needs to think short term. You need to think of projects that you can do quickly and that will look exciting to a search committee. A permanent job changes your perspective. You can finally think about what *you* want to do.

When I got to Cardiff I knew I didn't want to work on some intricate piece of the physics of extragalactic radio sources (Chapter 2), which might interest, if I was lucky, 100 people around the world. I wanted to do something bigger. When I looked back at the history of astronomy, most of the really big discoveries had been made with ingenious new instruments, especially ones that broke the ground in some new electromagnetic waveband. I began to think again about sub-mm astronomy.

I had also heard by then about a new sub-mm instrument that was being assembled at the Royal Observatory in Edinburgh. I had left sub-mm astronomy because it wasn't possible to take pictures. There were no sub-mm cameras. This new instrument was a camera, although only just. The camera in a mobile phone contains over a million pixels. The new camera, the Submillimetre Common User Bolometer Array (SCUBA), would have only 37. But it was a camera.

SCUBA would be the first-ever camera in a waveband in which the universe had barely been explored. It was obvious there was the opportunity for a big discovery. My idea was to use SCUBA to try to answer one of the biggest questions in my own research field: How were the elliptical galaxies formed?

<hr />

* The other is *Science*.

Galaxies fall into three main classes. The elliptical galaxies are the homely members of the family. There are always more pictures of the beautiful people, and it's the same way for galaxies. The NASA folk who run the *Hubble* website are obviously not great admirers because although there are plenty of pictures of the elegant spirals and vivacious irregular galaxies, I couldn't find a single picture of an elliptical galaxy. The picture of an elliptical galaxy that I have shown in Figure 4.1 was taken with a telescope on the ground. But because I, too, can't entirely resist a beautiful galaxy, I have also shown a picture there of the spiral galaxy Messier 51.

The beauty of a spiral galaxy is the result of a mix of dust and massive stars. The massive stars are the ones that emit most of a galaxy's light. A massive star may have a mass that's only ten times that of the Sun, but it emits *ten thousand* times more light. Although there are fewer of these massive stars, which are blue in colour, than the low-mass stars like the Sun, it is the light from the massive stars that usually dominates the overall light from a galaxy. But without the dust, a spiral galaxy would be a boring monochrome blue. It is the dust, which like smoke makes the light redder in some places and cuts it out completely in others, that makes the galaxy so beautiful. It is because there are so few massive stars in them, and so little dust, that elliptical galaxies like M87 are a little drab.

The origin of galaxies is not a complete mystery because galaxies are still being formed today. A massive star may emit 10,000 times as much light as the Sun, but this can't go on for long; the star is so prodigal with its nuclear fuel that its life will be 1,000 times shorter than the Sun's. Their very short lives mean that we can use the number of massive stars in a galaxy to make an estimate of the recent stellar birthrate. In our galaxy, a typical spiral, the best estimate is that a mass of gas equivalent to about three Suns is turned into stars every year. This may not seem very much – there are roughly 300 billion stars in the Galaxy – but over time it mounts up. Over the lifetime of the universe (14 billion years), even a stellar birthrate this low would be enough to make most of the stars in the Galaxy. Galaxy-building is still happening today even if very slowly.

And sometimes it is happening very fast. Figure 4.2 shows two of the celebrities of the galaxy world: the Antennae galaxies. These are two galaxies that have strayed so close to each other they are being ripped apart by each other's gravity. They can no longer escape from each other – like two feuding lead actors cast in the same movie – and they will continue orbiting each other, stretching and ripping chunks off each other, until – computer simulations suggest – they eventually merge, building one big galaxy out of two smaller ones.

The origin of elliptical galaxies, though, is more of a mystery because in them galaxy-building seems to have stopped completely. They contain very few massive stars, which shows that the birth of stars has virtually stopped in these galaxies. Most of their stars have very low masses and are very, very old, with an age of about ten billion years, almost as old as the universe itself. The stellar birthrate is low because these galaxies contain very little gas and dust, the raw material out of which stars are made. Astronomers, who are as anthropocentric as the next human, often call ellipticals 'dead galaxies' because there is so little happening. They may not be very glamorous but some of them are very big. The biggest ellipticals contain over ten trillion stars.

FIGURE 4.1 At the top the elliptical galaxy Messier 87 (M87), at the bottom the spiral galaxy Messier 51 (M51).

Credit: Canada-France-Hawaii Telescope, J.-C. Cuillandre CFHT) and NASA, ESA, S. Beckwith (STScI) and the *Hubble* Heritage Team.

FIGURE 4.2   Galaxy superstars – the Antennae galaxies.

Credit: NASA, ESA and the *Hubble* Heritage Team (STScI/AURA)

Not much may be happening now, but we can deduce something important about the birth of an elliptical galaxy. Since the stars in a present-day elliptical are all roughly the same age,[†] about ten billion years, all these stars must all have been born at roughly the same time. The stellar birthrate in an elliptical today may be very low, but ten billion years ago the galaxy must have been popping out stars. To make all the stars in one of the biggest ellipticals, the stellar birthrate ten billion years ago must have been between 100 and 1,000 times higher than in our own galaxy today.

Ten billion years ago, a big elliptical would therefore have been much more glamorous. Along with the low-mass stars that are still present because of their long lives, there would have been many massive stars born at the same time, which we don't see today because of their spendthrift expense of nuclear fuel. To explain such a high stellar birthrate, the galaxy then must also have been full of gas. With so many massive stars pouring out light in all directions, the galaxy would have been a cosmic lighthouse, up to 1,000 times as luminous as any galaxy today.

[†]   It is possible to estimate the average age of the stars in a galaxy from its spectrum – another thing you can learn from spectroscopy.

Astronomers often use the term 'proto-elliptical' for an elliptical during this early star-forming period. Since it is the period in which most of the galaxy's stars were born, it makes sense to think of this as the time of the birth of the galaxy itself.

But surely this is all speculation? This did happen 10 billion years ago. Luckily for astronomers like me, it is not the *past* that is dead and buried.

The past is in front of our eyes. If I look at the Sun, I am not seeing it *now*. I am seeing what the Sun looked like 8 minutes and 17 seconds ago, which is the time it has taken its light to reach me. I am therefore looking back in time, although it seems safe to assume that not much has changed about the Sun during the last eight minutes. The closest star (other than the Sun) is the star Proxima Centauri, which is at a distance of 4.2 light-years. It takes the light from Proxima Centauri 4.2 years to reach us, and so we are now looking further back in time, *seeing* the star 4.2 years ago. Within the Galaxy, the time delays are not large enough for much to have changed; it is possible that Proxima Centauri does not exist anymore although it probably does because it's a feeble little thing that hordes its nuclear fuel and should outlast the Sun.

Outside the Galaxy, for a cosmic historian the time delays become more interesting. The Andromeda galaxy, the closest big galaxy, is about two million light-years away, which means we are *seeing* it as it was two million years ago when the first humanoids were roaming the earth. The closest cluster of galaxies, the Virgo Cluster, is 65 million light-years away, which means we are now looking back in time to when the dinosaurs were wiped out by an asteroid. The only limit to how far back we can look in time is that in the first 400,000 years after the big bang the universe was full of hot ionised gas, which is opaque to electromagnetic radiation, so we can observe objects all the way back to then but no further.[‡]

Ironically, this means it is actually the *present* that is dead and buried. We are *seeing* the Sun, Proxima Centauri and the galaxies in the Virgo cluster in the past, but because of the speed of light we have no way of seeing what they look like *now*.

The arguments I gave above suggest that a present-day elliptical galaxy was formed – born – about ten billion years ago. Even as I write this, it seems almost unbelievable to me that we can do this, but, in principle, we can look back in time and see the birth of a galaxy as it happens.[§]

There are some practical challenges, of course. We may be able to look back in time, but we also need to be able to tell how far back in time we are looking. As so often in astronomy, the technique we can use to do this is spectroscopy.

In the time between radiation leaving a distant galaxy and reaching our telescopes, the universe has expanded. The effect of this expansion is to stretch the radiation, increasing its wavelength by the same factor the universe itself has expanded. This increase moves the wavelengths of the spectral lines towards the red end of the spectrum, which is why we talk about the 'redshift' of a galaxy. We can measure its redshift from the change in the

---

[‡]   The cosmic microwave background radiation is radiation from this hot gas, so using this radiation we are able to observe the universe 13.7 billion years ago, only 400,000 years after the big bang (see my book *Origins*).

[§]   We can't see the birth of our own galaxy, of course. We can only see the birth of a galaxy that is far enough away that its light takes just the correct amount of time to reach us.

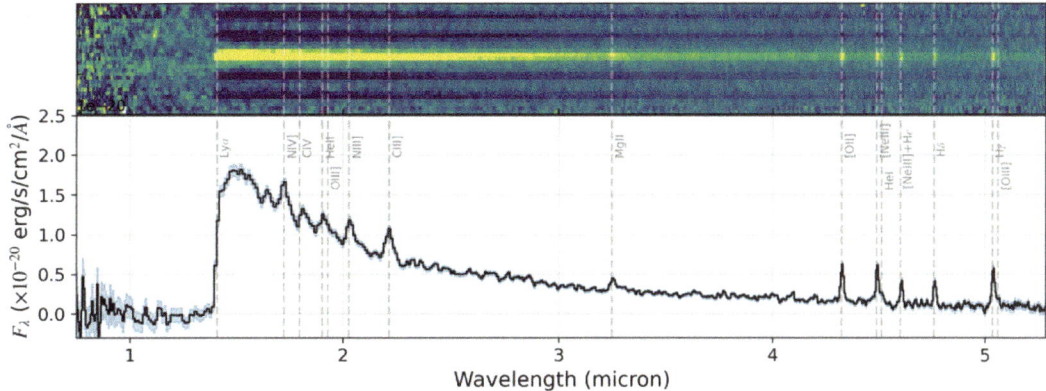

FIGURE 4.3   A spectrum of a galaxy obtained with the *James Webb Space Telescope*. The spectral lines are shown by the vertical dashed lines.

Credit: NASA, ESA and the JADES team

wavelengths of its spectral lines, which can then be turned into a time with a cosmological model, a set of equations that describe how the universe has expanded since the big bang. Therefore, if I have a spectrum of a galaxy, I can calculate how far back in time I am looking.

The redshifts can be immense. Figure 4.3 shows a spectrum of a galaxy with one of the largest redshifts known, which was recently obtained with the *James Webb Space Telescope*. The lines in this spectrum have been shifted in wavelength by a colossal factor of 11.6034, which means that in the time the radiation has been travelling, the universe has expanded over ten times in size. Using the standard cosmological model, I calculate that the radiation detected by *Webb* has been travelling for 13.3 billion years. We are therefore looking back in time 13.3 billion years, observing the galaxy directly – not inferring what it was like – 440 million years after the big bang. (This is another of the things I find almost unbelievable.)

The other practical thing we need is a big enough telescope. The faintest galaxies that have been discovered by astronomers are about ten billion times fainter than the faintest star in the night sky, which makes it challenging to obtain a picture of one of them let alone get a spectrum.⁵ If you start on a career as a cosmic historian because of a passion to understand the formation and evolution of galaxies, good luck to you! Your dedication and brilliance may overcome everything, but your success is likely to depend on your access to big telescopes, which will depend as much on luck, the places you get postdocs and the connections you can make at the major observatories, as your ability as a scientist. The bigger

---

⁵   In spectroscopy, the light is divided into lots of little wavelength chunks, each of which is recorded by a separate detector. The light falling on a detector is therefore much less (and the signal-to-noise ratio less – see Chapter 2) than the light falling on one of the detectors in a camera.

the mirror of your telescope mirror, the more light it will collect – and the further back in time you will be able to see.

In the early 90s, cosmic historians got a big boost (at least those at the right universities) with the opening of the first of the Keck Telescopes on Mauna Kea, which were then and still are the biggest telescopes in the world, with mirrors ten metres in diameter. With so much collecting area, it became possible for the first time to measure redshifts for galaxies ten billion years in the past – the birth time of ellipticals. Packed full of massive stars, proto-ellipticals should have been cosmic lighthouses, easy to find.

Except that they didn't seem to be there. When astronomers used Keck and other large telescopes to look for the proto-ellipticals, they didn't find them.

The arguments that they should exist are straightforward and no more complicated than the way I have described it in this book. There was only one explanation that anyone could think of for the absence of the proto-ellipticals: dust. With such a high stellar birthrate, it seemed likely that a proto-elliptical would also have contained a very large amount of gas, and where there is gas in the universe there is usually also dust. It seemed likely – nobody could think of a better idea – that it was this dust that was absorbing the visible light, shrouding the proto-ellipticals from the view of telescopes that observe in the visible waveband, even ones as large as Keck.

But is never possible to hide the energy in radiation completely. If it is missing in one waveband, it must have been transformed – laundered – into another waveband. It seemed likely that if the proto-ellipticals were missing in the visible waveband, the visible light had been absorbed by the dust, which must be re-radiating the absorbed energy in another waveband (otherwise the dust would keep on getting hotter - Chapter 2). It was easy enough to show that the most likely waveband was the sub-mm waveband. If the proto-ellipticals were not present in images in the visible waveband, it should be possible to see them in sub-mm images.

And there we were, in the mid-90s, on the verge of getting the world's first sub-mm camera. The person who could get their hands on SCUBA first might discover the proto-ellipticals.

That was my big idea. Unfortunately, many other astronomers had the same one.

Astronomers often work in teams. This is sometimes because it is more fun (although sometimes it's the opposite), but it is mostly because when a new instrument comes along, especially one that seems likely to make some important discovery, it makes sense to work together. If you and 30 other astronomers submit separate proposals for observing time, only one or two of you will be successful. If you apply as a team, you stand a much higher chance of being part of the discovery even if it means you will need to share the credit. When SCUBA came along, most of the UK astronomers interested in searching for the missing proto-ellipticals got together and formed a single team.

I didn't join them. One of the downsides of working in a team is that most of the important scientific papers end up being written by the leaders of the team and their postdocs. I didn't have a postdoc and I had only just returned to the UK, so nobody knew me very well, and I could see nobody was going to let me write one of the main papers.

I decided to start my own team. I asked Simon Lilly and Dick Bond, two colleagues from Toronto, to join me, which I thought was a good move because Canada also had observing time on the James Clerk Maxwell Telescope. To give our team some technical credibility, I asked Walter Gear, who was the leader of the team that was building SCUBA, to join our team. I knew Walter slightly and he was quite happy to join us.

It seemed unlikely that both teams' proposals would be accepted. I heard from a source within the other team (big teams leak) that they were not happy about the upcoming competition. I also heard from the same source that there was already squabbling within the team, which contained some big egos. Walter was also part of the other team. One of the big egos told Walter that he couldn't be in both teams. He must choose between us. Walter told me it didn't take him more than a second – he chose us.

Since everyone who mattered in the UK sub-mm community was in the other team, I didn't have a lot of confidence that our proposal would be accepted. I decided I needed a fall-back plan, a really good idea that would get me some observing time on SCUBA if the proto-elliptical idea fell through. One day in my office I had one.

SCUBA was such a primitive camera – 37 pixels!** – that it would only be able to take pictures of tiny areas of sky. For the proto-elliptical search, this wouldn't matter. The proto-ellipticals were so far away (far away, far back in time – same thing) that even a sub-mm image of a small area should find a lot of them – in the same way that the tall buildings in a city centre appear very close together if you see them from a distance.

The Hubble Deep Field is a nice example of the pluses and minuses of trying to learn about the universe from images of a tiny area of sky. This is a book about sub-mm astronomy, but *Hubble*, in most ways an old-fashioned optical telescope, keeps elbowing its way in. The reason it is impossible to ignore is that its launch in the early 90s started a revolution in astronomy, allowing astronomers to study the universe in much more detail than had previously been possible.

Our images are always blurred, whatever the telescope, even if the beautiful *Hubble* images don't always look it. Any image made with electromagnetic radiation is blurred because waves travel around corners. Any wave encountering an obstacle – an ocean wave passing through a gap in a sea wall, light passing into a telescope's aperture – is bent by the obstacle, which in a telescope has the effect of blurring its images. The blurring depends on both the diameter of the mirror or lens and the wavelength of the radiation. There is less blurring with a larger mirror or a shorter wavelength. In Chapter 3, I made a big pitch for the importance of sub-mm astronomy, but I didn't mention the one huge disadvantage. The ugly truth of our waveband is that because sub-mm radiation has such long wavelengths, our pictures are badly blurred. A picture taken with a telescope in the visible waveband will always show much more detail than a picture taken with a sub-mm telescope of the same size. A complication in the visible waveband is that there is additional blurring in this waveband from the effect of the atmosphere. The importance of *Hubble* is that it is an optical telescope, with its intrinsic ability to see fine detail, but without this additional blurring.

** There was also a second array of 91 detectors, but it wasn't used much because it used one of the flakier atmospheric 'windows'.

The *Hubble* headquarters is the Space Telescope Science Institute in Baltimore, and I was lucky to be there during the exciting period just before the launch, although I missed the launch itself, which was after we moved to Canada. (I did get the chance, though, to see the astronauts who were going to take *Hubble* up in the shuttle when they visited the Institute. They seemed surprisingly old, but that was probably because almost everyone in the Institute at the time was under 40).

It was the director of the Institute, Bob Williams, who was the inventor of the Hubble Deep Field. Directors of telescopes usually get a small percentage of the observing time on the telescope. Bob decided to blow all his observing time on very long exposures through four different filters on one tiny patch of sky – the most sensitive optical images ever made. There are now even deeper images, made with *Hubble* itself and with the *James Webb Space Telescope*, but it is still one of Bob's pictures that is the iconic one (Figure 4.4), the one on many astronomer's walls.

The images cover a tiny area of sky, about the area covered by a small coin at a distance of 20 metres. The brilliance of Bob's idea is that the Hubble Deep Field is effectively a bore-hole through time. An analogy I like is the cores through the ice cap that scientists use to study the changes in the Earth's atmosphere; the ice at different depths has been laid down at different times and will contain gas bubbles from the atmosphere at that time. The Hubble Deep Field has given us a core through the universe. There are galaxies in the picture from every period, from galaxies that we are seeing only half a billion years in the past to ones we are seeing almost 12 billion years ago, only a couple of billion years after the big bang. If you were allowed to only use one set of images to learn about the evolution of galaxies, this would be a good set to choose. By comparing the properties of the galaxies in the picture seen at different times, you would immediately discover a lot about the changes in the galaxy population over the last 12 billion years.

That is the plus of images of a small area of sky. The big minus is that if you relied on the Hubble Deep Field to learn about the universe, you would miss a lot of things. You would never know about galaxy clusters because the area of sky covered by the Hubble Deep Field is much smaller than even a single cluster. You would also never know about many kinds of rare objects because there often aren't any examples in such a small area. When the first radio astronomers discovered radio galaxies and quasars, they did it by a radio survey covering most of the sky (Chapter 2), not by an ultra-sensitive survey of a tiny area.

This was the same problem we would have with SCUBA. The sub-mm waveband was a new one. Everyone knew that we might make some big discoveries. And the biggest would not be proto-ellipticals, which we had already guessed were out there, but the beasts that nobody knew existed, like radio galaxies and quasars had been for the first radio astronomers. But this would require a survey of a large area of sky, and with SCUBA, a camera with so few pixels, this wouldn't be possible.

Then one day in my office I had an idea. I realised we might not be able to discover truly unimaginable beasts – like radio galaxies and quasars had been for the radio astronomers – but we would be able to look for some more imaginable beasts: galaxies containing ultra-cold dust.

FIGURE 4.4   The Hubble Deep Field.

Credit: R. Williams (STScI), the *Hubble* Deep Field Team and NASA

We already knew that galaxies contain a lot of dust from the survey by the Infrared Astronomy Satellite (IRAS) in the 80s (Chapter 2). IRAS had surveyed the sky in the far-infrared waveband, so it was only able to detect relatively warm dust (the warmer the dust, the shorter the wavelength of the radiation it emits – Chapter 2), and it would have missed any galaxies that contain only cold dust. Cold dust emits sub-mm radiation. I realised that if we used SCUBA to observe enough nearby galaxies, we might detect one that contained some of this very cold dust. The discovery of a galaxy containing ultra-cold dust would not be as exciting as the discovery of radio galaxies and quasars, but it would still be something new. I talked to a few other people about this idea and put together a team.

Somebody came up with a name – the SCUBA Local Universe and Galaxy Survey – and I submitted the proposal.

The day of the time allocation committee meeting is always stressful. I knew this one, which was going to be in Edinburgh, would be particularly stressful because it was when the decisions would be made about all the SCUBA proposals. It didn't help that I was on the committee myself. I had a sleepless night in an Edinburgh bed and breakfast before the meeting and spent the following day trying to calmly discuss other people's proposals while waiting for my own to be discussed. I had to wait in the corridor while the other members of the committee discussed my proposals, so by the end of the day I still didn't know whether we had any observing time.

I got the emails a few days later. I had been awarded observing time for both projects, both for the search for ultra-cold dust in nearby galaxies and the search for proto-ellipticals. But I also heard, through my usual source, that the other proto-elliptical team had been awarded time as well.

Collaborators are often necessary in a science project, somebody with essential skills that you don't have, for example. If you are lucky – I generally have been – the collaborators also make the project more fun. And if you are very lucky, the collaborator's own ideas, intellectual insight and critical ability – all the things that make science a creative process – enhance the project. I have had a few collaborators like this. Loretta Dunne was one of them.

I met Loretta for the first time when she was an undergraduate in Cardiff and chose one of my research projects. A year later, she was looking for something to do after her degree and I was looking for a PhD student to work with me on the SCUBA data. At the time, she was more interested in a career as a pilot, but I managed to persuade her that she could still do that after she had done a PhD with me. We agreed that the survey of nearby galaxies would be her PhD project, although she would also work with me on the search for proto-ellipticals.

I had been awarded a lot of observing time with SCUBA, but I was giving a lecture course and I had other teaching commitments, so I realised it wouldn't be possible for me to go on every observing run. The rules said that a new PhD student should only go on an observing run with an experienced observer, but Loretta seemed very competent. I found someone at the observatory who said they would keep an eye on her, and only a month after she started her PhD, she flew off to Hawaii for her first observing run.

This was the first observing run for the nearby galaxy project. Loretta had a list of galaxies that we knew, from previous IRAS observations, contained a large amount of dust. We planned to observe all these galaxies with SCUBA to find out if they contained any of the ultra-cold dust. Loretta spent the first observing run of her PhD using the new camera to take the first-ever pictures of galaxies in a new waveband.

A month later, it was my turn when I went on an observing run for the proto-elliptical project. We had a slice of luck because our observing time was in December, while the other team's was not until January. At the beginning of December, Loretta and I flew out to Hawaii. When we got to the Big Island we met Simon Lilly and drove up the mountain.

When we arrived at Hale Pohaku we looked for Wayne Holland, the SCUBA instrument scientist. It was usually a pleasure to chat with Wayne, a warm funny guy from Manchester. This time when we found him he looked very serious. He was really apologetic. SCUBA was broken. He said it would take him 14 days to warm SCUBA up to room temperature, open the cryostat, fix the problem and then cool everything down again. Unfortunately, this would be four days after the end of our observing run. We had a dismal lunch and started off back home on the 24-hour journey.

That Christmas I was a bit distracted. I knew the other team's first observing run was scheduled for January. I also knew there was an El Niño weather complex over the Pacific, which only happens roughly every ten years and usually creates exceptionally good sub-mm observing conditions. Our own second observing run was scheduled for February.

Sometime that February I was waiting on St. Mary Street in Cardiff at three in the morning, among the drunks spilling out of the clubs, for the coach to Heathrow. I didn't know what had happened on the other team's observing run the previous month; I could have asked my source but I didn't really want to know. Four hours later, on a bleak, wet winter morning, I got out at Heathrow. As I retrieved my bag from the coach I heard my name on the airport public address system. The message asked me to call my wife. As the rain dripped down around me, I stood at a payphone at the side of the road and called Keirsten. She told me that Wayne had called from Hawaii. SCUBA was broken again. Wayne said not to come out.

The following morning back in Cardiff, I checked my email. There was a message from Wayne. He had made a mistake. SCUBA was not broken after all.

It would have been a lot less funny if none of us had managed to get there, but Loretta and Simon had gone out to Hawaii the day before me and were now at Hale Pohaku. While I was stuck in Cardiff, Loretta and Simon sat on the summit of Mauna Kea and started the observations. They didn't have time while they were observing to look at the results, so a week later, on an overcast afternoon, I sat down and started to run the computer programs that would turn the raw data recorded by SCUBA into a sub-mm picture of the sky.

When the picture appeared on the screen, I wasn't sure what I was looking at. I had never seen a sub-mm picture before and it didn't look like any picture I had ever seen. There was nothing in it that looked like a galaxy. There were just two patches, which might be sub-mm sources but might not. I wasn't sure.

But we had observed this area of sky for a reason. A few years before, Simon and a team of French and Canadian astronomers had observed the galaxies in this area in a survey in the visible waveband. When I looked at their paper, I found they had discovered a galaxy right at the position of one of the patches in the SCUBA picture. The patch in the SCUBA picture must be a sub-mm source, and the galaxy in Simon's paper must be the one emitting the sub-mm radiation. On a grey February afternoon in Cardiff, I was looking at one of the first ever pictures in the sub-mm waveband.[1] I had discovered that a galaxy billions of years ago was emitting strong sub-mm radiation.

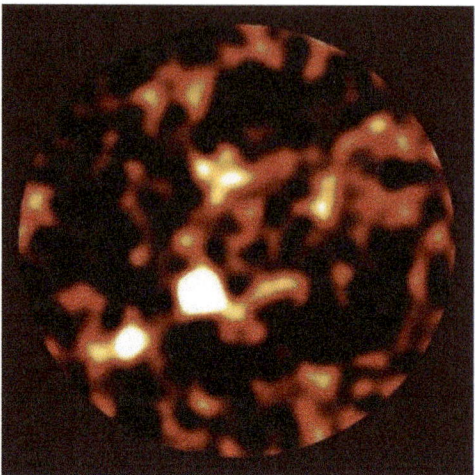

FIGURE 4.5 The SCUBA image of the Hubble Deep Field.[2] The white patches, 'blobs', are the sub-mm sources.

Credit: The James Clerk Maxwell Telescope

We had lost. It turned out that the other team's observing run in January had been successful. They had used SCUBA to observe the Hubble Deep Field itself, eventually building up a total exposure time of 50 hours on this one tiny patch of sky.

Their sub-mm picture[2] (Figure 4.5) is nowhere near as pretty as the optical picture (Figure 4.4), but I think it's probably more important. The handful of white patches show the places where SCUBA detected sub-mm radiation – sub-mm sources. Sub-mm pictures don't show much detail, so each source appears as an undifferentiated 'blob'. When the team looked in the *Hubble* image at the position of the brightest sub-mm source, they found there was nothing there. The team realised that the sub-mm radiation must be from a galaxy that is so shrouded in dust that it is virtually invisible. There is so much dust in these galaxies, which everyone started calling the 'sub-mm galaxies', that initially nobody was able to measure their redshifts. When this was eventually done almost a decade later,[3] it turned out that most of the redshifts are between 2 and 3, which means we are observing these galaxies between 10.4 and 11.6 billion years in the past. With a redshift, it's straightforward to calculate the luminosity of the galaxy, which is typically between 100 to 1,000 times greater than the luminosity of any galaxy today.

Everyone realised immediately that the sources detected in the first SCUBA images, the sub-mm galaxies, must be the missing proto-ellipticals. It turned out, though, that both teams had been scooped. The first people to see the sub-mm galaxies were three young astronomers: Ian Smail, Rob Ivison and Andrew Blain. They used SCUBA to observe some nearby clusters, with the idea of using the gravitational fields of the clusters as giant lenses to magnify the sub-mm radiation from more distant galaxies (Chapter 11). They saw the sub-mm galaxies before the rest of us,[4] but everyone who saw these sources in their images quickly reached the same conclusion: these were the missing proto-ellipticals.

The sub-mm galaxies are everywhere. They are so bright they turn up in even short exposures with sub-mm cameras. In any sub-mm picture – of a nearby galaxy, a filament of dust in interstellar space, a planetary system – there will probably be some sub-mm galaxies lurking in the background. We still believe, 25 years after they were discovered, that these galaxies are the ancestors, ten billion years ago, of the elliptical galaxies we see around us today[5].

While I had been focussing on the search for the proto-ellipticals, Loretta had been dealing with the data from the ultra-cold dust project. By the end of her PhD, she had sub-mm pictures of almost 200 nearby galaxies, the first ever pictures of galaxies in the sub-mm waveband. She had not discovered any galaxies with only cold dust, but she had discovered that galaxies contain much more cold dust than the warm dust detected by the earlier IRAS observations.[6] She was awarded a prize by the Royal Astronomical Society for her PhD thesis.

During her PhD we had many meetings in my office, usually sitting around the whiteboard so we could sketch on it to try to make sense of our results. Sometimes during one of these meetings, the conversation would turn to what we really wanted to do. When you open a new waveband, what you should really do is survey a large area of sky to find the new kinds of objects that nobody knows are out there. What we really wanted to do was a sub-mm survey of a large area of sky. But with a camera that only had 37 pixels, it just wasn't possible.

It was all a little frustrating.

CHAPTER 5

The Makers

O<small>N A DISMAL</small> D<small>ECEMBER</small> morning in 2018, long after the demise of *Herschel*, I am in a taxi on the way to the European Space Research and Technology Centre (ESTEC*). ESTEC, the nerve centre of the European Space Agency, is just outside the little seaside town of Noordwijk in the Netherlands. I am on my way to interview Göran Pilbratt, the *Herschel* Project Scientist, to find out what really happened.

Through the taxi's windows all I can see are vast flat brown fields stretching to the horizon, drainage channels ruler-straight, plastic poly-tunnels, the occasional windmill, and every now and then, in this densely settled land, a jumble of neat Dutch houses. When we eventually reach ESTEC, a complex of low undistinguished buildings seemingly plonked down at random in this flat landscape, it all seems rather anonymous; apart from the sign at the entrance I might be at the headquarters of a large insurance company.[†]

Inside, immediately after the reception area, is a sprawling, confusing canteen/conference area, which I later discover was designed to encourage scientific collaboration. It doesn't look much different to me from my university cafeteria back home: people working on laptops, small groups drinking coffee. The only obvious differences are that everyone looks about ten years older than the students back home and they are all wearing security badges. At ESA, they clearly take security seriously. I don't like wearing anything around my neck and generally try to lose a security badge as quickly as possible, but I decide not to lose this one. I feel there would be consequences.

Eventually, away from this central area in a maze of corridors, I find Göran's office. Here the differences from my university are more obvious. Everything is eerily quiet – no students. The walls are covered with pictures of past space missions and there is the occasional model of some spacecraft. Göran's office, however, is like any university professor's: untidy, uncarpeted, cheap furniture with a view over those flat Dutch fields. Göran himself is Swedish in his sixties with a booming laugh. He has spent 30 years in the Netherlands, speaks Dutch but still supports Sweden when the two countries play football.

---

[*]   Someone forgot about the R.

[†]   It was designed by a famous architect to blend in with the sand dunes, apparently.

 DOI: 10.1201/9781003195290-5

The role of the Project Scientist is something I always found difficult to understand. For me, and astronomers like me, Göran was the face of *Herschel*, the person who made the pronouncements, who told us what we could and couldn't do with the telescope – the Man. But behind him were the people we never saw, the people who really were in charge: the Project Manager, who led the team that built the telescope, and the Mission Manager, who was in charge of the mission after launch. The Project Scientist in any big science project is the liaison between the scientists and the engineers. The Project Scientist takes the wish list from the scientists, talks to the engineers, finds out what's practical, goes back to the scientists, sees what they think about this and so on. Göran did not build or design any part of *Herschel* himself, but he worked alongside the engineers who did. He didn't sign any cheques, he couldn't tell anyone what to do, he wasn't even very high in the ESA hierarchy, but he was still one of the most important people in the *Herschel* team. Soft power, I guess you'd call it.

Göran became *Herschel* Study Scientist in 1991. He became Project Scientist in 1995. Today, over two decades later, he spends a small part of his time dealing with the aftermath of the mission and the rest of it planning a new space mission, which will be launched long after he retires. This is typical of ESTEC where the goal is not to educate or do research but to design cool space missions, which other people will then use to do their research. Everything is driven by the missions and ESTEC staff like Göran are assigned to a mission and then moved abruptly when their job there is done.

The career of Stephan Ott, the leader of the team that wrote the software to process the *Herschel* data, is typical. While still a student Stephan was talent-spotted by ESA and hired to work on the software for *Huygens*, the European part of the joint US-European *Cassini* mission to Saturn. When *Cassini* arrived in the vicinity of Titan, Saturn's largest moon and the only moon in the solar system with an atmosphere, *Huygens* separated from *Cassini* and parachuted down to the surface, the first time a spacecraft had landed on the moon of another planet. Descending through an atmosphere of methane and nitrogen, *Huygens* landed in one of the weirdest places in the solar system, a surface covered with lakes of liquid methane and ethane. But by this time Stephan was long gone. He had finished his part of the project, so his bosses at ESTEC had moved him to work on the software for the *Infrared Space Observatory*. And then to *Herschel*, and after that to *ExoMars*, the ESA mission designed to answer the question of whether there has ever been life on Mars.

For Göran, *Herschel* was therefore old news. But as we talked during a long afternoon, as dusk descended over the flat Dutch landscape and he began to remember incidents from *Herschel's* turbulent story, he got more excited, his booming laugh more frequent….

A space mission never has a single creator. It is always an ensemble effort, with a team of thousands, which changes constantly as team members leave to do other things and new members join. The team that finishes the mission, when the fuel or the funding runs out (or when the spacecraft explodes or breaks down), is often completely different from the team that started the mission decades before. A space mission is a little like a Hollywood movie that has been rewritten so many times and had so many changes of director that nobody remembers who originally came up with the idea. But if one did have to name the person who was the ultimate source of the river of money, people, high-tech machinery and ideas that eventually became *Herschel*, it would have to be Thijs de Graauw.

Thijs was part of a group of young astronomers who in the 70s began to explore the universe in a new way. This astronomy new wave began when carbon monoxide molecules, a carbon and oxygen atom bound together, were discovered in the Orion Nebula.[1] We see the nebula (below Orion's belt) because it contains very hot gas which emits light. The carbon monoxide (CO) in interstellar space is very cold and completely invisible, but it emits bright spectral lines in the radio waveband, which were detected with one of the new radio-telescopes. The discovery of CO showed that there must be much more invisible cold gas around the nebula than the very hot gas we can see. Molecular hydrogen, two hydrogen atoms spliced together, doesn't emit bright spectral lines,[‡] but there should be a lot more of it than CO – there are more than 10,000 times the number of hydrogen atoms than carbon atoms – so the detection of CO showed there must also be a huge amount of molecular hydrogen. The Orion Nebula is therefore the tip of an iceberg. It is the tiny, hot visible part of a huge cloud of cold molecular gas, enough gas to make 100,000 stars like the Sun – the Orion giant molecular cloud.

Thijs and the other new-wave astronomers became fascinated by this new way of doing astronomy. They began to detect many other chemical compounds in space and became captivated by their bizarre chemistry. Apart from molecular hydrogen which dominates everything (there are 10,000 times more hydrogen molecules than CO molecules), here is a list of 19 of the 20 most common molecules in interstellar space, starting with the most common and working down the list: carbon monoxide, molecular nitrogen, hydrogen deuteride, the hydroxyl radical, the trihydrogen cation, molecular oxygen, the methylidyne radical, the cyanogen radical, diatomic carbon, the ethynyl radical, hydrogen cyanide, formaldehyde, hydrogen isocyanide, ammonia, the formyl radical, tricarbon, carbon monosulphide, sulphur monoxide and methanol.[2] A few are familiar but most do not exist on earth because they would react instantaneously with the other gases in the atmosphere; it is only deep in interstellar space where a molecule rarely encounters another molecule that they can exist at all. The other thing that fascinated everyone at the time (and still does) is how many of the molecules contain carbon, the element whose properties make life possible. It seems quite possible that life on Earth is somehow linked to the strange chemistry of molecular clouds (Chapter 14).

The molecule I have missed off the list is water because it has been hard to find out how much water exists in the universe. There is a lot on our planet, and since it is made of two of the most common elements in the universe, hydrogen and oxygen, there should be a lot in the universe at large. But it's been remarkably difficult to find out exactly how much. One problem is that a lot of it is locked up in ice, which is hard to detect. The other is our location.

[‡] Molecular hydrogen does emit spectral lines but only when it is heated to a few 1,000 kelvins, well above the typical temperatures of interstellar gas.

We live on the blue planet, which overall is a good thing (if there was no water we wouldn't be here), but it does create a big problem for astronomers because the water vapour in the atmosphere absorbs most of the sub-mm radiation from space. The sub-mm waveband is also the waveband in which the water molecule emits its strongest spectral lines.[§] Back in the 70s, the surface of a watery planet like the Earth turned out to be the worst place to be if one wanted to find out how much water exists out there in the universe.

To get above all this water, Thijs and his colleagues began to think seriously about a sub-mm telescope in space. Thijs was a member of the panel NASA set up to discuss the idea. NASA ultimately rejected it, but Thijs persevered. By then he had a job at ESTEC. He organised a workshop to discuss the idea, which took place in May 1982. In that November, he and a small group of European scientists put in a formal proposal to ESA to launch a telescope that would observe in the sub-mm and far-infrared wavebands.

In normal times the proposal would have been read by some ESA committee and quietly forgotten.[3] Most ideas for space missions never see the dark of space, and ESA really had no money. But these were not normal times. A year later, Roger Bonnet worked his magic and conjured up both an exciting new space programme and the money to make it happen (Chapter 1). In 1984, the scientists meeting in Venice to draw up the programme for Horizon 2000 considered the proposal of Thijs' group and selected the *Far-Infrared and Submillimetre Space Telescope* (*FIRST*) as one of the four Cornerstone missions. With an X-ray telescope at the other end of the electromagnetic spectrum, a mission to study the Sun and the Earth's magnetic field and a mission to make the first landing on a comet, Horizon 2000 was about as broad a programme as it was possible to get. The scientists were enthusiastic, Bonnet coaxed the politicians into providing the money, and Europe finally had a space programme to rival NASA's.

Göran's story of how he became involved in *FIRST* started with a girl. As a teenager in Sweden, he was interested in a girl who lived along his route to school. As they walked to school together, sometimes they talked about what they would do with their lives. Göran told the girl he was interested in astronomy, but becoming a professional astronomer was obviously impossible so he was going to become a pilot instead. A few years later, he started his Swedish military service and was accepted for flying school. He would need to spend the first month crawling around in forests with the other pilots-to-be, but then he would learn to fly.

After two months of flying over the Swedish forests, his instructor took him aside, told him he was not cut out for flying and grounded him. But he still had to finish his military service, so he was sent to another Air Force base, where he spent his time making coffee for the real pilots. There he met a guy in the photo lab who happened to be a student at Chalmers University. The guy said that for anyone interested in a physics degree there

---

[§]   It's not a coincidence. There is a law of physics, Kirchoff's Law, that states that a gas that emits radiation strongly at a particular wavelength will also absorb radiation strongly at the same wavelength.

was only one choice: engineering physics at Chalmers. Göran did the degree. Then, since Chalmers happened to have a radio observatory, a PhD in radio astronomy.

Göran ended up at ESTEC again because of someone he met: a guy from Stockholm called Urban Frisk who worked at ESTEC and told him what it was like there. When Göran finished his PhD he was offered postdocs at two very good places, but he decided he really wanted to work at ESTEC. He phoned up Urban and explained the situation. He quickly got an invitation to come to ESTEC for an interview. He got the job.

ESTEC is all about the missions. The study scientist is the person assigned by ESA to study a potential space mission. When Göran arrived at ESTEC in late 1986, his friend Urban was the study scientist for two possible space telescopes: a radio-telescope called *QUASAT* and another telescope called *FIRST*. He tried to interest Göran in both missions. Göran went to a few meetings about *FIRST* but thought it was a little boring. *QUASAT* seemed more exciting and in line with his radio astronomy expertise.

A few years later, Urban left ESTEC, and ESA needed a new study scientist for *FIRST*. They chose Göran. At ESTEC you change missions when your boss tells you to. In the spring of 1991 he became the study scientist for *FIRST*. (*QUASAT* was never launched.)

He joined *FIRST* at the beginning of the time of crisis. Seven years on from when it was written down on a piece of paper in Venice, not much had happened. *FIRST* was stuck in Phase A although an industrial study of the whole project was about to be completed. Phase A in ESA mission-speak is when the scientists and the engineers, working together, decide on the science requirements of the mission, and whether the mission is feasible given the technological and financial constraints. If the answer to the question is no, little money has been spent and it's easy enough to call the whole thing off; many missions never emerge from Phase A. It is still easy for ESA to cancel missions in early Phase B (Phase B1). And it is only in later phases (Phases B2, C and D), when stuff is being built and real money spent, that it gets difficult to cancel the mission. (Phase E is launch and operations, so then it really is too late.)

When Göran joined the team, he learned there would soon be a big opportunity. A mission becomes a lot safer, whatever its formal status, once it has appeared in a launch schedule. The first two Cornerstones of Horizon 2000 already had launch slots. The first, *SOHO/Cluster*, was due to be launched (with unfortunate consequences – Chapter 1) in 1996. The second, the X-ray telescope *XMM-Newton*, would be launched in 1999. The third and fourth Cornerstones were *FIRST* and *Rosetta*, which would be the first spacecraft to land on a comet. No decision had yet been made about which mission would go first. ESA announced that the third launch slot was now up for grabs. It was a big deal for both teams, not only for the reason above but also because any space mission takes such a large chunk of a scientist's career that any chance to speed things up is always welcome.

The challenge for the *FIRST* team was that in going for the third launch slot they might still end up with their mission cancelled. The big problem, seven years after Venice, was it still wasn't clear whether the mission was possible. When the results of the industrial design study came in, the predicted cost of the mission was almost 100 million euros more than the money ESA had allocated. A shortfall of 100 million euros would probably be chicken feed for NASA; the soft rain of space money falling on their states mean that the politicians are always amenable to an appeal to increase the budget. But it was a huge deal for ESA. Bonnet had only been able to persuade the European politicians to fund Horizon 2000 by promising that all the missions would be completed within a fixed amount of money (Chapter 1). If one of the missions in the programme went drastically over budget, other missions would need to be scaled back or cancelled. And to the accountants and managers, the solution to a problem like that is obvious.

When Göran came on the scene, the most important player in *FIRST* was Reinhard Genzel, the chair of the *FIRST* Science Advisory Group. For the last 30 years, Reinhard, a hard-driving, creative scientist,⸙ has been Director of the Max Planck Institute for Extraterrestrial Physics near Munich. Reinhard would be a powerful personality in any walk of life, but in the hierarchical German system the word of a Max Planck director is law, so it must have been galling for him dealing with the engineers. Every time he visited ESTEC, they took something away from him.

In the plan sketched out in Venice, *FIRST* had a primary mirror eight metres in diameter. A mirror this big would not fit within the fairing of one of ESA's Ariane rockets, but in the scientists' vision the telescope's mirror would unfold once it was in space, like the petals of a shiny flower. The engineers got rid of this. Much too fanciful – unproven technology.⸬ They insisted the mirror be a traditional mirror, which meant it needed to be reduced to 4.5 metres. Then they realised even this was too expensive. The size of the mirror went down again. At one point there was even talk of the mirror being reduced to two metres.[4] For a telescope the size of the mirror is everything (Chapter 4), and so this was all pretty depressing – astronomers began to lose interest and leave the project.[5] Eventually, it was decided that a mirror three metres in diameter would be possible.⸬⸬

Everyone knew there was no chance that *FIRST* would get the third launch slot, and might even be cancelled, unless the mission got back on budget. This meant a downscope, the ugly piece of space jargon that everyone hates, although the engineers hate it a little less than the astronomers because it makes their life easier. The astronomers have been promised a telescope that will be able to do all these wonderful things. Then suddenly someone is saying, sorry, we won't be able to do all the things on your list. Too much money, too difficult. One by one, things are thrown off. That is the downscope.

This is one of the reasons for the Project Scientist. In a downscope, the Project Scientist is the voice of the astronomers who are going to be using the telescope. The Project Scientist tells the engineers which are the things on the list that the astronomers really want and should not be thrown off in any circumstances while telling the astronomers which things are too technically difficult and add so much to the budget they are no longer worth doing. The critical part of the downscope is the negotiations between the Project Scientist and Project Manager, trying to keep as many good things on the list for the astronomers but trying to make life as easy as possible for the engineers. For over a decade, the Project

---

⸙   He won the Nobel Prize in 2020 for his investigation of the super-massive black hole at the centre of our galaxy.

⸬   A primary mirror that unfolds like the petals of a flower was eventually used, 30 years later, for the *James Webb Space Telescope*.

⸬⸬   The mirror's diameter was eventually increased to 3.5 metres, which was the largest size that would fit within the fairing of the new Ariane 5 rocket.

Manager for *FIRST* was Thomas Passvogel. In the ESA hierarchy, the Project Manager and Project Scientist are deliberately made to report to different bosses to ensure that the Project Scientist is independent of the project team. Thomas and Göran got on very well, going together on visits to some of the industrial companies that were building parts of the observatory and spacecraft, and enjoying a beer together and dinner. There was even talk at one point of Göran becoming a formal member of the project team, but ultimately it was thought that the independence of the Project Scientist was too important a thing to lose.

At this moment, everyone accepted that a downscope was inevitable. A not quite so wonderful telescope in space is better than a wonderful telescope on paper. Over a period of 18 months, a small group of scientists and engineers who became known as the *FIRST* Tiger Team made a frantic effort to redesign the mission.

The reduction in the size of the primary mirror helped a bit, but it was still not enough, so the team changed other parts of the design. In the original plan the telescope's instruments, its cameras and spectrometers, were to have been cooled with a huge tank of liquid helium, but the cost of launching something this large into space added a lot to the budget. They replaced the helium tank with a complicated set of mechanical coolers. The loss of the helium tank meant that it was no longer possible to cool the instruments to below 4 kelvins, which was not low enough for the cameras that would observe interstellar dust. And so these went as well. *FIRST* became a telescope that would only be able to observe molecular gas. At this, there was such an outcry from the astronomers that the team put the cameras back, although nobody was quite sure how to make them cold enough. Even with this cost-cutting, *FIRST* was still over budget, but the team decided it had gone as far as it could.

The other thing needed to win the competition with *Rosetta* was a strong science case. Reinhard thought this would be easy. He accepted that comets, which have been in deep freeze for four-and-a-half billion years, must hold clues about the origin of the solar system, but he thought the instruments on *Rosetta* would be too simple to tell us much about this. The first landing on a comet would be exciting – in a Thunderbirds kind of way – but he didn't think it could compare with discovering how the stars and galaxies were born. One of the important moments during the development of any space mission is the production of the mission's Red Book,[‡‡] which describes the scientific programme that will be carried out by the mission. In the months before the meeting of the ESA Science Programme Committee, which would make the decision, the team wrote the *FIRST* Red Book,[6] which hits all the high spots of sub-mm astronomy: interstellar dust, molecular gas, the birth of stars and planetary systems, the birth of galaxies.

The fourth of November 1993 was decision day. Armed with the Red Book and a design that was at least not crazily over budget, Reinhard and Göran flew off to Paris for the meeting. Some members of the *FIRST* team thought they still stood little chance, there were still so many embarrassing holes in the design.[7] But after his presentation to the committee, Reinhard, who is hyper-competitive (according to somebody close to him: "If it's throwing

---

[‡‡] Back in the days when these books were printed, they always had red covers, but why this was even Göran doesn't know.

stones in the water, Reinhard will want to throw the biggest stone the longest distance"), was sure he had convinced them. They were scientists, after all. *FIRST* would do much more exciting science than the simplistic stuff possible from landing on a comet. He left Paris confident of the result.[8]

Rosetta became the third Cornerstone.

The explanation given by the committee was the length of its journey. It seemed likely that *FIRST*, even starting much later, would still make its first observations before *Rosetta* reached its comet.[§§] Everyone suspected the real reason was the holes in the design. The only consolation was it seemed clear that *FIRST* would have the final Cornerstone launch slot; the threat of cancellation had vanished-for the moment.

The next few years were spent filling holes. The ESA engineers came up with a design for a cooling system that would get the cameras down to the necessary temperature. It was a little complicated, consisting of 13 separate mechanical coolers, eight for regular operations and five spares, but the engineers thought it would do the job.

One of the biggest holes – a pretty deep one considering its importance – is that nobody had yet figured out how to make the primary mirror. Making the mirror for a telescope on the ground is difficult enough – smoother than silk and with precisely the right shape – and there are often a few failed attempts, but it is much harder to make the mirror for a space telescope. The mirror also needs to be strong enough to survive the juddering of launch, and for a sub-mm telescope like *FIRST* there is the additional requirement that the mirror must emit very little sub-mm radiation. The original plan was that the mirror for *FIRST* would be made from carbon fibres, but when this was tried it was found that while the mirror was light and strong, its surface wasn't smooth enough.

And the skeleton at the back of everyone's mind was what happened in 1991. 'First light', the moment when the first pictures are taken with a telescope, is usually a tense but ultimately happy occasion, but this wasn't the case for *Hubble*. Everyone now takes for granted the beautiful pictures it has produced, but when the scientists back on Earth saw the first *Hubble* images they were badly out of focus. The manufacturers had ground the mirror into the wrong shape, trusting the results from an expensive measuring machine which had been assembled incorrectly – and what was worse, ignoring the results from some simpler machines which had shown correctly that the mirror had the wrong shape. NASA had therefore just spent the US taxpayer's money – five billion dollars of it – on sending a telescope into space that was no better than one on the ground. The *Hubble* team was very lucky because their telescope was in a low-Earth orbit, so it was possible to send up a shuttle with a new instrument to correct the distortion. *FIRST*, though, would be too far away. The second Lagrangian point is roughly a million miles from the Earth, four times as far as the Moon. If anything went wrong there, that would be it. Nobody wants to be Project Scientist when something like that happens.

[§§] *Rosetta* reached Comet 67P/Churyumov-Gerasimenko in 2014, five years after the launch of *Herschel*. Its lander module, *Philae*, landed on the comet on 12 November 2014. I happened to be at ESTEC at the time, so I was able to watch the landing there. Reinhard was wrong. It was pretty cool.

The engineers eventually found a material that they thought could be used to make the mirror: gold-coated Carbon Fibre Reinforced Polymer (CFRP for short). Nobody had made a mirror this large out of CFRP yet, but the industrial company awarded the contract to make the mirror was confident it would be possible. There was a plan.

Then in 1996, *FIRST* was hit by a perfect storm. In June, the Ariane 5 carrying the first Cornerstone, *Cluster*, exploded shortly after launch (Chapter 1), and ESA had to scramble to find money for a replacement.

In July, ESA approved a new mission, *Cobras-Samba* (later renamed *Planck*). *Cobras-Samba* was not a Cornerstone mission but it would still cost half of a Cornerstone, putting even more pressure on the ESA budget. There are two big tribes of scientists in the ESA universe: there are the ones who are interested in the solar system – the places a spacecraft can visit – and there are the rest of us. Many of the solar system tribe, who had supported *Rosetta* and who would probably blow the entire ESA budget on exploring Mars if you gave them the chance, thought *Cobras-Samba* was just another sub-mm mission like *FIRST*,⁑ which given the strain on the budget meant both missions immediately came under threat (Chapter 1).

In November, a second mission, *Mars-96*, failed at launch. This was a Russian mission, but several European countries had contributed to the payload, and after a lot of pressure ESA agreed to pay for a replacement,⁂ which blew the budget completely.

And then a tornado blew in from the States. The NASA Administrator Dan Goldin proposed that ESA take a share in the *Next Generation Space Telescope* (*NGST*), the successor to *Hubble* that eventually became the *James Webb Space Telescope*, which was launched in 2021 (which just shows how long space missions really take). At that time the *NGST* was still in the planning stage. Goldin offered ESA a great deal: 250 million euros for a 50% share in the telescope. Everyone immediately fell in love with the idea. After its repair, *Hubble* was starting to take the amazing pictures that have changed the way we look at the universe. The astronomy community, still mostly optical astronomers who didn't really care about exotic missions like *FIRST* and *Planck*, loved the idea of being part of *Hubble's* successor. But even with Goldin's great offer, it was pretty obvious the only way ESA could afford to take it was by cancelling something. And given all the failures and new missions, it couldn't just be a small mission. They had to cancel something big.

The obvious target was *FIRST*.

The complex organisation of a space agency, a network of committees, managers and directors – a little like the nervous system of an octopus in which there are decision-making neural hubs in the tentacles as well as the brain††† – means it is often hard to work out how any decision was made, but as the result of some intensive investigative journalism⁹

⁑ The goal of *Cobras-Samba* (aka *Planck*) was to observe the cosmic microwave background radiation, in order to investigate the properties of the universe as a whole. One of the many goals of *FIRST* was to understand the origin of galaxies. Very different things!

⁂ This became the highly successful *Mars Express*, which took pictures of the Mars surface with such detail that we would have been able to see trucks moving across the surface if any existed.

††† See *Other Minds* by Peter Godfrey-Smith.

I can state fairly confidently that the fate of *FIRST* was decided at a meeting of the ESA Space Science Advisory Committee.

Its name does not make it sound particularly important, but the committee is there to advise the ESA Director of Science on which space missions to support. The main topic of this meeting, which took place sometime in 1997, was to advise the Director of Science on whether to accept Goldin's offer. If the committee decided to recommend acceptance, it would not necessarily mean the end of *FIRST* – Reinhard Genzel was the chair of another key ESA committee (another neural hub) and the Director of Science, Roger Bonnet himself, was known to hate joint ESA-NASA projects because it made them vulnerable to cancellation by Congress – but the mission would be heading towards a cliff.

I never anticipated, when I started my career, how many days I would spend in committee meetings. But all those windowless rooms and bad coffee have taught me a few things. Rule number one is always turn up. However boring the agenda and tempting it is to skip the meeting, if you don't turn up there is a good chance that when you look at the minutes later you will find some decision you don't like.

Rule number two is that there are usually only a few people who care about any decision; everyone else is usually looking out of the window (or would be if there was one). The two people on the Space Science Advisory Committee who really cared about the decision about NGST were Riccardo Giacconi and Michael Rowan-Robinson. Riccardo, a pioneering X-ray astronomer,[‡‡‡] had been the director of the Space Telescope Science Institute, the home institute of *Hubble*, before returning to Europe to become director of the European Southern Observatory, the big European optical observatory. Everything in his career made him a natural supporter of the *NGST*. Everyone expected Riccardo would make a strong pitch for accepting Goldin's offer. On the other side was Michael Rowan-Robinson. Michael, a member of the *FIRST* team, had written part of the Red Book. He knew the possible consequences of accepting Goldin's offer.

The chair of the committee was the Dutch astronomer Lodewijk Woltjer, who had decided to hold the meeting close to his home in Geneva. As a relaxing social event, he arranged a cocktail party in his flat the evening before the meeting.

Michael knew the importance of the decision. During Lodewijk's party, he tried to talk to every member of the committee. He made a few basic points. ESA is also part of *Hubble* but might as well not be for all the credit it gets. ESA pays 15% of the budget but everyone thinks of *Hubble* as an entirely American mission. He argued that *NGST* would be just the same; Europe would pay a huge amount of money but would be steamrollered again by the NASA publicity machine. He also argued that NASA's estimate of the cost of NGST was ludicrously low. *NGST* was likely to cost much more than 500 million euros. If it did, ESA would have to match NASA to keep their equal share or become a minor partner – like *Hubble* again.[§§§]

Riccardo broke my number one rule for committees. He didn't turn up. At the very last moment, he decided not to come and sent a substitute. Michael made the same arguments

---

[‡‡‡] He won the Nobel Prize in 2002.
[§§§] He was right. The cost of the *James Webb Space Telescope* has been north of ten billion dollars.

in the meeting he had made the night before, and rule number three is that good arguments do usually win. Without Riccardo present to argue the case, the committee decided to recommend to the ESA Science Director not to accept Goldin's offer.

FIRST survived the storm and all that remained now were the technical challenges. Sometime in 1996, *FIRST* got a new Project Manager, Hans Steinz. Before *FIRST*, he had been the Project Manager of the *Infrared Space Observatory*, on which the instruments had been cooled with a tank of liquid helium. Hans swept away the baroque monstrosity of 13 separate coolers and brought back the big tank.

Also in that year the industrial company given the contract to make the primary mirror out of CFRP said that, sorry, but it would need more time. ESA revoked their contract. The *FIRST* team then looked at two other possibilities: aluminium or silicon carbide. Göran thought the second was the better bet, but the Project Manager disagreed – and on a space project the Project Manager is the boss. A French company, Aerospatiale, said they had made a similar aluminium mirror for a military satellite and, pas de problème, it would be trivial to do something similar for *FIRST* (I picture them shrugging and clicking their fingers). A year later they had failed. CFRP then came back into the picture. NASA said that they could make a mirror out of CFRP using a different process. As everyone knows the Americans are the masters of high technology, ESA expressed a collective sigh of relief and agreed to let NASA take care of it. Three years later, the Americans had failed to make a mirror, and it was now getting worryingly close to the launch, which was then scheduled for 2007. Fortunately, at the time of choosing aluminium, ESA, to avoid putting all their eggs in one basket, had also started a backup project to investigate whether it was possible to make a mirror out of silicon carbide. This did turn out to be possible, which is a fitting end to the story because the eventual mirror (Figure 5.1) was made from the same material as interstellar dust (Chapter 3).

By the turn of the millennium, *FIRST* was safe. The money was beginning to flow and it would have been hard to cancel. Göran decided that such a marvellous telescope needed a better name than a jumble of letters. The year 2000 would be the bicentenary of William Herschel's discovery of infrared rays (Chapter 2). To Göran, his seemed the perfect name for the telescope: the discoverer of infrared radiation, which was roughly what the telescope would observe[†††]; an astronomer born in one European country but who worked in another – useful for a European team with so many touchy nationalities; and one of the greatest astronomers of all time. The correct procedure would have been for Göran to put his idea in a memo, give it to his boss, who would then forward it up the ESA chain of command. Instead, Göran decided to send a memo directly to the ESA Director of Science, Roger Bonnet. In his memo, Göran suggested that Bonnet announce the name at a meeting arranged to discuss all the wonderful science that *FIRST* would do, due to be held in Toledo, Spain, in December 2000.

[†††] The telescope would observe radiation in the far-infrared and sub-mm wavebands, whereas William Herschel discovered near-infrared radiation (Chapter 2). But the importance of William's experiment is that it showed there was radiation outside the visible waveband, so in my opinion the name is still a good one.

FIGURE 5.1 *Herschel* being inspected at ESTEC. The primary mirror is the thing with a hole in its centre. The shiny structure at the left is the telescope's sunshade. The metal struts attached to the primary mirror support the smaller secondary mirror. The telescope was designed so that radiation from a distant galaxy would be reflected by the primary towards the secondary, then reflected by the secondary through the hole to the instruments, the cameras and spectrometers, in the cryostat below the primary mirror.

Credit: ESA

He didn't hear anything for months. Bonnet was so far above him in the hierarchy he rarely saw him, but at ESTEC one day he saw him in the queue in the canteen. He hesitated, but decided it was now or never. He asked Bonnet in his direct way why he hadn't replied to his memo. Stung by the question, Bonnet testily responded that he hadn't replied because he didn't like the suggestion. Göran got the feeling he had never read the memo. A few months later, Bonnet, who was scheduled to give a welcome talk at the Toledo meeting, phoned up Göran to discuss what he should say. After discussing his talk for 30 minutes, he finished by saying that he liked Göran's idea of renaming *FIRST* – and the *Herschel Space Observatory* was born.

All it needed now were some instruments.

Matt Griffin became the leader of the team that built one of *Herschel's* instruments because his boss didn't like going to meetings. Peter Ade, the head of the sub-mm instrumentation group at Queen Mary College, had been one of the pioneers of sub-mm astronomy (Chapter 2).When the round of meetings to discuss the instruments began, he would have been a natural choice to lead one of the teams, but Peter loved to spend time in his lab and resented any time away from it. He needed someone to represent his group. He sent Matt.

The meetings to discuss the instruments for *Planck* were happening at the same time. The plan was that there would be two instruments for *Planck* and three for *Herschel*, and since the technology was quite similar – and to save money – ESA sorted out both instrument programmes together. By the end of all the meetings and all the horse-trading, what happened was what always happens in the ESA world: each of the big countries got to lead one but only one instrument team. The UK ended up as the lead country for one of the three instruments on *Herschel*: the Spectral and Photometric Imaging Receiver (SPIRE**** for short), an instrument that combined a camera and a spectrometer.

And Matt became the team's Principal Investigator, the PI. It was a popular choice. Everyone knew that Matt had all the technical skills; he had been part of a team that had built a similar instrument for the *Infrared Space Observatory*. And in a community full of academic prima-donnas and scientists who would struggle to organise a jumble sale, there are not many of us capable of leading a huge, complicated international technical project. Matt, by common consent, was one of that tiny handful. Everyone trusted that with him as PI, SPIRE would be a success.

Anyone, though, who knew much about sub-mm instrumentation wasn't quite so sure. The big challenge facing all the *Herschel* instrument makers was that much of the technology didn't yet exist. This is not a problem for anyone who wants to build an optical camera for a space telescope. There are still big challenges, but at least there have been hundreds of similar cameras made for telescopes on the ground – the basic technology is well understood. However, when Matt and his team started to design SPIRE, their only model was SCUBA, the world's first and so far only sub-mm camera (Chapter 4). SCUBA had been a huge success, but with an array of only 37 detectors, compared to the millions of detectors in a modern phone, it was hardly even a camera. The Red Book had promised that *Herschel* would do a sub-mm survey that would cover a large area of the sky. Matt knew that this wouldn't be possible with a camera with so few detectors. His team had to build something more ambitious.

And then there are the problems faced by anyone building an instrument that is going to be launched into space. The instrument needs to (a) survive in a radiation-soaked environment; (b) satisfy all the spacecraft's power, thermal and space constraints; (c) be operated remotely from Earth; (d) never, ever need servicing or fixing; and (e) survive being shaken on top of one of the most powerful machines ever built by humans.

With any space project, there is also a big bet you need to place. Space projects take so long that the technology available when the project starts is often much more primitive than the technology available by launch. At some point you therefore need to make a choice: between technology that you know will work but may be outdated by launch and taking a gamble on cool new, unproven technology, which you hope you can get working by launch day. And hope is an essential part of a project like this. At the beginning,

---

**** Yes, it doesn't make sense – it only works by the E being the second letter of the final word.

there are usually parts of the instrument that nobody quite knows how to make. Everyone crosses their fingers and hopes that somebody, eventually, will figure out how to do it.

Something that did make Matt feel hopeful was his team. He had worked with many of them before on the instrument for the *Infrared Space Observatory*. One of the members of the previous team Matt most wanted for the new team was Bruce Swinyard, who worked at the Rutherford Appleton Laboratory in Oxfordshire. One day in Matt's office at Queen Mary College to discuss some business from their previous instrument, Bruce said he would really like to work with Matt on the new *Herschel* project. With Bruce in the team, Matt felt a lot happier.

The SPIRE team consisted eventually of 150 scientists from 8 countries, but from the beginning Matt and Bruce were its creative heart, working closely together on all aspects of the design and development of the instrument. They argued a lot, with a lot of swearing from Bruce, but they thought this was a strength. If there are two possible solutions to a problem, one way to settle it is for each person in the partnership to take one side and argue it out. One thing they did agree about was that they would forget about ego. They would not aim at building an instrument to impress other instrument scientists. They would build the instrument the astronomers wanted.

In my limited experience, a space project seems to progress from crisis to crisis: the budget is cut, the spacecraft engineers change the mass and power limits for the instrument, some international partner wants to pull out, there is a technical problem nobody had anticipated. Whatever the crisis, Matt and Bruce's attitude, which they didn't realise until they thought about it later, was that they always believed there was a solution. They would figure something out. Whatever the obstacle in the path of the project, they would 'bulldoze through it, burrow under it or circle around it'.

In the beginning was the drawing.

At the beginning of any creative project, there must be a moment when something is written down. A project only really starts once there is something tangible to work on.

For an instrument, that moment is the conceptual design. Matt knew that the astronomers wanted to survey as big an area of sky as possible, so he needed to make the detector arrays as big as possible. He also thought it was important that SPIRE be able to take pictures at several wavelengths simultaneously. In his original sketch (Figure 5.2) – the foundation document for SPIRE – the instrument would contain three arrays of detectors, each array detecting radiation in a different wavelength range. In the sketch, the radiation collected by the mirror enters from the left, is divided into the different wavelength ranges by dichroic mirrors (the dotted lines) and then detected by the three detector arrays (the grey rectangles). There is no detail in the diagram. Matt didn't know how many detectors he would be able to squeeze into each array, what they would be made of or much of anything else. It was just a sketch.

Bruce's conceptual design (Figure 5.3) showed how the rest of the instrument might fit around the components in Matt's sketch (which in Bruce's diagram are all in one of the grey boxes). His design was important because it was the first attempt to sketch the cooling system. The SPIRE detectors would need to be cooled down to about 0.3 kelvins, even colder than the 1.65 kelvins of the helium in *Herschel's* big tank. SPIRE would need its own

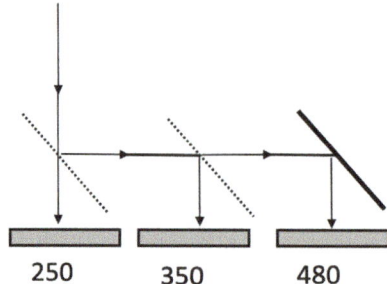

FIGURE 5.2   Matt's initial conceptual design for the camera part of SPIRE. The dotted lines are dichroic mirrors, which separate radiation into different wavelength ranges. The sloping sold line is a regular mirror and the three grey boxes are the detector arrays.

Credit: Matt Griffin

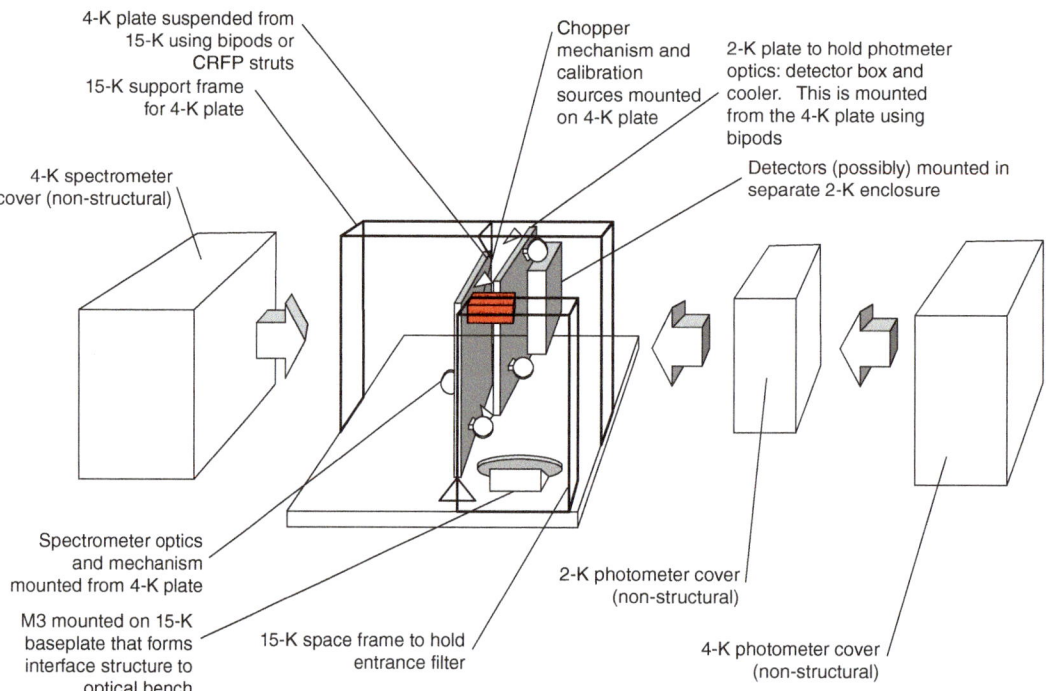

FIGURE 5.3   Bruce's conceptual design. The details are not important, but the diagram shows Bruce's original vision of how the different parts of SPIRE might fit together. In this drawing, everything in Matt's sketch is in one of the smaller grey boxes in the centre.

Credit: Matt Griffin

additional cooling system to cool the detectors down that final bit. One of the team's biggest challenges was figuring out how to do this.

The conceptual design is only the first step. Each of those empty boxes needs to be filled with a real mechanism. Somebody needs to design the mechanism, and somebody needs to make it and get it to work. All these parts then need to be assembled into a single instrument. In any creative work, the medium of the art form always sets some fundamental limits. These limits for SPIRE were set by the spacecraft engineers, who laid down how much electrical power it was allowed to use, how much space it could use, and, critical for something that is going to be launched into space, its maximum mass.

But this all makes it sound much too easy. I have never built an instrument myself, but from talking to those who have, it is clear this is not a linear process. The design of every part of SPIRE impacted the design of other parts, with a new design of one part of the instrument often forcing the redesign of other bits of the instrument. The creative souls who designed and made the parts of SPIRE were never working in isolation – otherwise all the parts, when they were assembled, would never have worked together. The conceptual design may have been the beginning, but most of the creative process was in making a real instrument from these rough sketches.

Sometimes, in this process, some of the original ideas were lost. The original plan had been that SPIRE would include a spectrometer that used a grating like an optical spectrometer (Chapter 3). But gratings that work at sub-mm wavelengths are very bulky. There was a detailed design of the spectrometer, but the team eventually realised it would be too big given the space constraints and replaced it with a more compact kind that didn't need a grating. But there are always trade-offs. The new design included a mirror that would be moved backward and forward with a spring, which is always dangerous in space; moving parts can break and there is no way to fix them.

Organising such a complex process must be very challenging. There is no management guide to running a big international space project. How do you run an organisation spread over eight different countries and when you can't give orders or fire anyone?

I suspect the leaders of most big international science projects end up discovering the technique used a lot by Matt. Let us call it the competition method of project management, although there is probably some fancy French sociological term. The basic idea is that if there is any contentious issue within the team, decide it with a competition. If the rules of the competition and the judging are fair, everyone will accept the result. At least that's the idea – and as leader of the project, if people don't accept the result, since you don't pay them, there is not much you can do about it.

For Matt, the biggest test of this method was over the detectors. At some point, the team had to place a bet on which detectors to use. There were three choices. The safe bet was the detectors used in SCUBA (Chapter 4). These were bolometers, essentially tiny thermometers. Their big disadvantage was that above the bolometers are little metal horns to funnel the radiation to the bolometers. Because of these horns, the detectors can't be packed together closely, which means that a picture taken with a camera has gaps because of the spaces between the detectors. It is possible to fill the gaps by combining pictures taken in slightly different directions, which was what was done with SCUBA, but it was not ideal. But there were now also two riskier bets, new types of detectors[††††] under development. These detectors could be packed together like the detectors in a phone camera. The two groups developing them claimed they would be ready by launch.

[††††] These were also bolometers but designed in such a way that they didn't need the metal horns. For clarity, I've just used the term bolometer for the detectors that did use the horns.

Given the magnitude of the project to build a space camera, which never costs less than tens of millions of euros, the contract to make its detectors is always a huge deal for a lab – money, jobs and a lot of prestige. The big problem for Matt was that the groups offering to provide the detectors were some of the most important ones in the SPIRE team. The bolometers were being proposed by a group at the Jet Propulsion Laboratory in the States, and the groups proposing to provide the new types of detectors were at the Goddard Space Flight Centre in the States and at the French Atomic Energy Commission in Paris. The proposal by the French group, which was led by Laurent Vigroux, was a particular problem because France was investing as much money in SPIRE as the UK. Given the way the world works, it was not unreasonable for Laurent to think that, as long as his detectors were as good as the other groups', his lab should get the contract. *Juste retour* is a French phrase, after all.

Matt decided to hold a competition. To be fair to the groups proposing the new types of detectors, he gave each of the groups a further 18 months to develop their detectors. The detectors would then be tested and technically assessed. He set up a panel of judges to make the final decision. To be as fair as possible, he included judges from the SPIRE team, from ESA, with some independent judges and a careful balance between nationalities.

As usual in the space world, nothing quite went to plan. The person given the job of testing the detectors was a postdoc at Queen Mary, Pete Hargrave. The plan was that Pete would fly out to the States, test the detectors made by the two US groups in their own labs, and then test the French detectors back in the lab at Queen Mary. But everyone's schedule slipped and all the tests had to be done very late in the day in London in a lab jammed full of equipment and people. In the middle of testing the JPL detectors, Pete got a phone call and had to rush home because his wife was in labour. Then, a week before the selection meeting, the French group claimed the tests of their detectors were being affected by electrical interference, which meant a metal cage had to be quickly built around the test system to shield it. The French were still unhappy with the results, so Pete, now a new father, flew over to Paris and spent two days testing the French detectors in the French group's own lab.

The final decision took place at the end of a two-day meeting in Building 67 at the Rutherford Appleton Laboratory. Matt, in his proper way, calls it the 'array selection meeting'; everyone else I interviewed calls it the 'shootout'.

The judging panel listened to presentations by the three groups. They looked carefully at the results from the tests. They balanced the promise of the new detectors against the possibility they might not be ready in time. Their vote was almost unanimous.

They made the safe bet and chose the bolometers.

The unsuccessful groups were not happy. If there was one time when France might have pulled out of the project, this was it. But Matt and Laurent Vigroux had a very good relationship‡‡‡‡ and Laurent thought it had been a fair competition. Even so, Matt had to spend an uncomfortable afternoon being interrogated about the decision in the basement of the French National Space Agency in Paris. But it was obvious that Matt had been

---

‡‡‡‡ Given the huge French investment in SPIRE, Laurent was co-PI of the project.

FIGURE 5.4 The winner of the *Herschel*-SPIRE Consortium Stakes, which took place at the Oxford Greyhound Stadium and was one of several races sponsored by the SPIRE team. The winner is being congratulated by several of the team members with Matt Griffin the tall one at the back.

Credit: Matt Griffin

as scrupulously fair as it was possible to be. France stayed in SPIRE[§§§§] and the project proceeded to its next crisis.

The other rule in the Griffin School of Project Management was to make the project fun. He tried to do this by scheduling meetings of the whole team every six months in each of the eight countries in turn, which he made sure always included an enjoyable social event. The Roman emperors gave the people bread and circuses, Matt gave them whisky tastings and dog races (Figure 5.4).

But none of what I have written so far seems to get to the essence, the soul, of the instrument. Let me make one final attempt by telling the story of one of its hundreds of parts.

This was a mechanism that was designed and built to solve one of the problems that everyone had known about from the beginning. The problem arose from the need to cool the detectors down to 0.3 kelvins. In the original plan, the detectors would be enclosed in a metal box attached to the tank of liquid helium. The box would therefore be at the temperature of the helium: 1.65 kelvins. For various reasons, the additional refrigeration

---

[§§§§] The detectors made by Laurent's group were eventually used for the other camera on *Herschel*, so everyone was happy.

system necessary to cool the detectors down to 0.3 kelvins needed to be outside the box. In the plan, this refrigeration system would be linked by a copper rod, which would pass through a hole in the box, to the detectors. Since copper conducts heat exceptionally well, the link would be enough to cool the detectors down to 0.3 kelvins.

The problem was that if the rod touched the metal of the box, the rod, and therefore the detectors, would quickly heat up to 1.65 kelvins, which doesn't sound very much but would be enough to make the detectors useless.

As an engineering problem, this doesn't sound too difficult: pass a copper rod through a hole in a box in a way that the rod doesn't touch the box itself. The problem was to ensure that the mechanism suspending the rod as it passed through the hole didn't itself transfer heat between the rod and the box. It also needed to be strong enough that the rod wouldn't be shaken free during launch.

Matt announced a competition. The winner of the competition to design the mechanism was his own postdoc, Pete Hargrave. Pete's idea was to suspend the rod in a network of Kevlar fibres, a synthetic material used in bulletproof vests. Kevlar would be ideal, he thought, because of its strength and because it conducts hardly any heat. As a material, though, it does have some challenges. If Kevlar fibres are bent through too great an angle, they lose their strength, so knots made from them are not very strong. Another challenge is that Kevlar is one of the few materials that expands when it is cooled, which will loosen the knots – a big deal in a camera that is being cooled down to 0.3 kelvins.

Pete solved these problems by designing a cat's cradle of fibres in which each fibre did not bend through too great an angle (Figure 5.5). In his design, the knots holding the fibres in place would not need to be too strong because the fibres would be wound around tiny capstans, which would reduce the strain on the knots. He also designed a system of spring washers to keep the fibres taut when the camera was cooled. The document in which he presented his design was 30 pages long with 27 separate figures – and this was for only one small part of the final instrument.

His prize was to make the mechanism. So one afternoon he sat down in the lab and assembled it. He was particularly careful in how he tied the knots, aware that if one of them came loose, the copper rod would touch the box, the detectors would warm up to 1.65 kelvins – and bye-bye SPIRE.

Ten years of evenings in the lab, hours stolen from partners and children, weekends spent trying to solve some critical problem, hundreds of reports and conference calls, thousands of emails – and SPIRE was finally as ready as it was ever going to be (Figure 5.6). It had been tested by every test in the imaginations of a lot of obsessively careful engineers. Each component had been tested many times in the lab. Then, once SPIRE was assembled, the engineers tested it at room temperature, cooled it down, tested it again at the temperature it would be in space, warmed it up… then did this all again. They also strapped it to a testbed and shook it around to simulate launch (the top of a seriously wonky washing machine on a high spin rate during an earthquake is the best analogy). Everything checked out. They had done every test they could think of.

Matt was sure it would work.

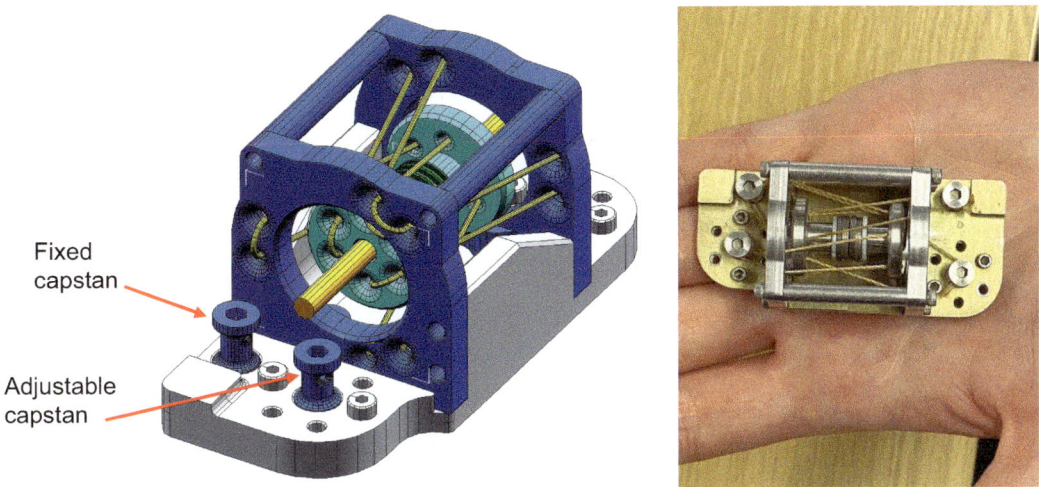

Fixed
capstan

Adjustable
capstan

FIGURE 5.5 *Left*: Pete's design. The copper rod (yellow) passes through the light blue structure, which is then held in place by the Kevlar fibres (yellow). The frame (dark blue) fits within the hole in the box. *Right:* the spare copy of his mechanism which is now on display in my department in Cardiff.

Credit: Peter Hargrave

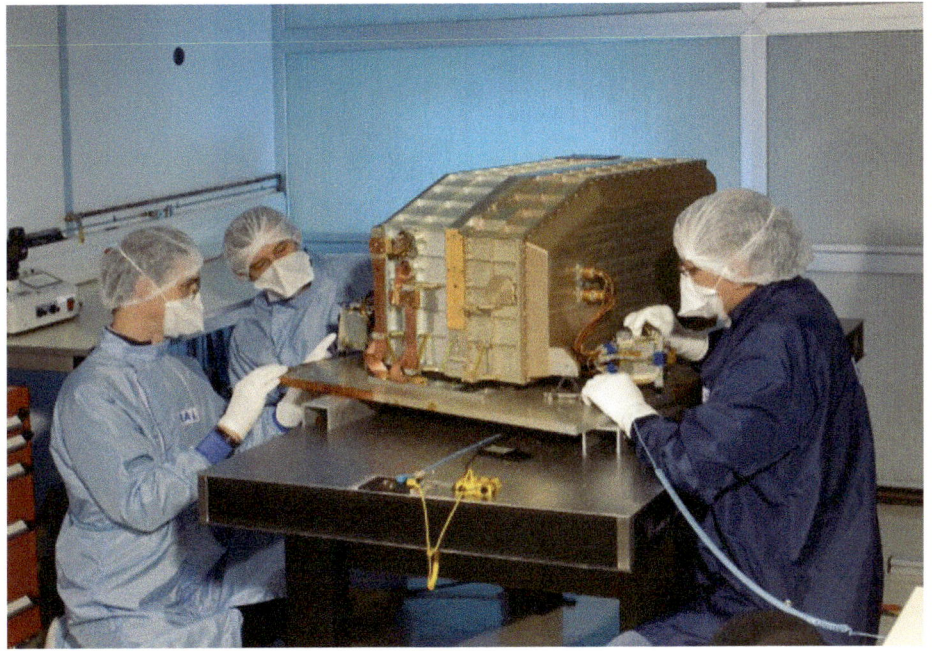

FIGURE 5.6 SPIRE in the lab.

Credit: Matt Griffin

It was not a soulless machine. Many parts of it had been assembled by hand. Every part had been thought about, worried over, maybe even dreamed about, by someone in the team. One part they had given a lot of thought was the spring that moved the mirror in the spectrometer, one of the few moving parts. The spring was flimsy because a moving mirror with a stiff spring uses too much electrical power. But the flimsiness meant the mirror needed to be locked in place during launch. At the last moment, one of the team decided the latch holding the mirror in place was too weak, and a box on the side of the spacecraft had to be quickly designed and built to hold the latch in position. Everyone had their own private worries about their own bit of the instrument. Matt was proud of his team. They had built the instrument the astronomers wanted. The members of his team had put big parts of their lives into it. They had done every test they could possibly think of.

He thought it would work.

CHAPTER 6

The Astronomers

I DON'T REMEMBER WHEN I heard about *Herschel* for the first time. It can't have been much before 2000, which means it must have been almost 15 years after it was picked as one of the ESA Cornerstone missions. Space astronomy had never interested me much. Space missions take a long time to plan and I have never been interested in telescopes I can't use immediately. There are some people who get involved in space missions from the start. They are prepared to spend years sitting in meetings planning future observing programmes that may never happen. For the sake of the community I am glad there are people like that. I am just not one of them.

It must have been around 2001 when *Herschel* first became more than a name. That was the year we heard the astronomy instrumentation group at Queen Mary College* was on the move. The group was building state-of-the-art equipment for two prestigious European space missions, *Herschel* and *Planck* – work which was being funded by huge grants from the UK government. But the group was getting very little support from their university. Big grants, in this case many millions of pounds, usually make university administrators come running, but Queen Mary was in the middle of an expensive merger with a medical school. The financial crisis meant there was no money for new positions, and every time an astronomer left the university the position was frozen. The group had had enough. They decided to try to move their entire group, people and equipment, to a different university. We heard we were on their list because Cardiff already had a few sub-mm astronomers. According to the rumour mill, the other places on the list were Cambridge and Sussex.

So one morning sometime in 2001 a coach from London full of scientists and engineers pulled up in our car park. We spent the day listening to talks from members of the two groups. Then the heads of the Queen Mary group, Peter Ade and Matt Griffin, went off to see the Cardiff vice chancellor, who promised them money and lab space (Cambridge, the rumours said, had offered them portacabins in the Cavendish car park). And then everyone in both groups went to the pub together. I could have written the script. The weather

---

* It was known at the time as Queen Mary and Westfield College. Its formal brand today is 'Queen Mary, University of London'.

DOI: 10.1201/9781003195290-6

was perfect, Cardiff looked beautiful in the sunshine, and when I looked around the pub there were mixtures of the two groups chatting and laughing. Then the Queen Mary coach had engine trouble, which meant they had to stay in the pub with us until late. By the end of that evening, I was fairly confident we were top of their list.

Soon after he moved to Cardiff, Matt Griffin asked me to join the SPIRE science team. Matt was opening the door on a big prize: the mission's guaranteed observing time.

I had always thought the guaranteed observing time is something that is fundamentally wrong with space astronomy. I thought this observing time was often used very inefficiently and the whole process grossly unfair. Observing time on telescopes on the ground is mostly 'open time', observing time that is available for anyone to use. It's often hard to get observing time – it's a fierce competition – but it is a fair one.[†] On a space telescope, however, a large part of the observing time, the Guaranteed Time (GT), is reserved for the scientists and engineers who built the telescope's instruments. There's a lot of it. Roughly one-third of the observing time on *Herschel* would be reserved for the instrument teams, which considering that the observatory ended up costing over one billion euros is equivalent to a staggering amount of money.

As an observer, I thought the idea that instrument builders would be able to make effective use of the GT was ridiculous. I have always suspected that Matt thinks that what I do as an observer is easy – just a matter of pointing a telescope at some interesting object in the sky. I believe that Matt thought designing the GT observing programme would just be a matter of reading the Red Book and choosing some obvious targets. I have spent my whole career cultivating my ability to design observational programmes – observations of carefully selected samples of objects that reveal something interesting about the universe. Nobody would think I, as an observer, could just walk into a lab and make an instrument. So why is it OK to think the other way around[‡]?

Of course, it's not quite as simple as that. Everyone knows that people like Matt may say they plan to use their instrument, but they never actually do it – once their instrument is finished they usually start thinking about the next one. The people who would make the most use of the GT were the astronomers that Matt had added to the team along the way, who were mostly astronomers at the same institutions as the SPIRE team or ones he knew from previous space missions. It had always struck me as unfair that a motley group of astronomers had access, just because they happened to work at the same institutions as the people who made the instruments or because they were mates of the PI, to millions of euros of observing time.

Now, for some reason, things felt a little different. It was me who had been asked to join the SPIRE team. Suddenly the GT system made a lot more sense.[§] I said yes immediately.

[†] I have simplified things a bit. On some telescopes some of the observing time is reserved for particular nationalities and some universities have their own private telescopes, but the statement is broadly true.

[‡] Matt claims none of this paragraph is true, but it is a fair reflection of how I, as an observer, thought about this at the time.

[§] I still think it is a bad system, although I can see why it exists. It exists because, in the ESA world, the spacecraft and telescope are paid for by ESA, but the instruments are paid for by national governments. The governments need to get something back for their money – hence the GT that goes to the instrument teams. A fairer system would be if the GT went to the whole national astronomy community rather than the rag-taggle of astronomers who happen to be at the same institutions as the instrument teams or are mates of the PI. More recent ESA space missions, such as *Euclid*, use something closer to this system.

The instrument teams had at their disposal about 10% of the total observing time[¶] – in money terms worth about 100 million euros.[**] Each of them organised their GT programme in a different way. Albrecht Poglitsch, the PI of one of the other instruments, decided just to divvy up the GT. He gave some to each of the institutes in his team in proportion to the contribution it had made to building the instrument, leaving it up to the astronomers at each institution how they used it.

Matt decided against this because it made it too difficult to carry out projects that required a lot of observing time, especially the big surveys promised in the Red Book, which he had in his mind from the moment he started designing SPIRE (Chapter 5). He decided to do things a different way.

Guided by the Red Book, *Herschel's* bible, he proposed to divide the GT observing programme into six research areas.[††] He said there would be separate groups in the team to plan the observing programmes in each area. He called these the 'specialist astronomy groups' but we all quickly forgot about this and started calling them the SAGs.

There were six of them. SAG1 was the birth and evolution of galaxies; SAG2 was nearby galaxies; SAG3 was the birth of stars; SAG4 was interstellar dust; SAG5 was the solar system; SAG6 included a hotchpotch of topics: supernovae, old stars and the origin of dust. Matt said every member of the team would be allowed to join up to two SAGs. I chose SAG1 and SAG2.

He didn't tell each SAG how much observing time it would get. Each SAG was asked to write an observing proposal, stating how much observing time it needed and how it would use it. A panel of senior members of the team[‡‡] would act as the time allocation committee and decide how much time to award each SAG.

It was a competition again but a pretty tame one. Anyone who wants a career as an observer gets used to rejection or changes careers very quickly. The chance of success of any observing proposal is small, from about 25% for a medium-sized telescope down to about 10% for the big beasts like the *James Webb Space Telescope* or the Atacama Large Millimeter Array. But this is open-time astronomy in which everyone is playing on the same pitch. For the people like me who had been tapped on the shoulder and asked to join one of the *Herschel* teams, this was an open goal; however bad our proposals, we knew we were all going to get something.[§§]

One of Matt's dafter ideas – at least I thought so – was that he had decided that each SAG would have two leaders. How do you settle matters if the two leaders don't agree about something? Matt in his careful way had spent a lot of time choosing the leaders so there

---

[¶]   The observing programmes drawn up by the instrument teams were subject to the approval of the Herschel Observing Time Allocation Committee – the HOTAC – but since these programmes were designed to fit precisely in their time allocation, the HOTAC's approval was pretty much a formality.

[**] I calculated this by assuming it is 10% of the total cost of the mission, which was about one billion euros. I sometimes wondered how much we would get if we posted the observing time on eBay.

[††] The decision about how to divide the observing time was taken by a committee, the SPIRE Steering Group. The proposal was Matt's, but the committee agreed it was a good idea.

[‡‡] The co-investigators on the original proposal to build SPIRE.

[§§] There is a rumour that one SAG's proposal was so bad that initially it didn't get any time, but the members of the SAG were made to rewrite it like naughty school children and were eventually given some.

was the perfect balance of nationalities and institutions.[⁵⁵] Given what happened later in some of the SAGS, I used to joke that perhaps he should also have thought about psychological compatibility. He never found this very funny.

In the case of SAG1, this wasn't an issue. The two leaders were Jamie Bock of the Jet Propulsion Laboratory in California and Seb Oliver of the University of Sussex. Jamie's group was working flat-out making the detectors for SPIRE (Chapter 5) and he was happy to let Seb, an easy-going guy, take the lead in designing the SAG1 science programme.

With the amiable Seb as our tour guide, in SAG1 we spent a couple of years having convivial meetings in beautiful cities around Europe and eventually decided to do what we knew we were going to do from the beginning. We would take some very deep sub-mm images of areas of sky for which there were already images in other wavebands.

The success of the Hubble Deep Field (Chapter 4) showed astronomers the benefits of spending most of their observing time on a few carefully chosen areas of sky. Images of fields like the Hubble Deep Field, with very long exposures, serve as windows into the universe, containing galaxies at every stage of cosmic history, from nearby galaxies, which we are seeing only yesterday in cosmic terms, to galaxies we are seeing only a couple of billion years after the big bang. I often describe the research I do as not very sophisticated because we can learn so much about the evolution of galaxies by a simple comparison of their appearances in images like these – galaxies seen at different times in the past. One of the big discoveries from the *Hubble* images, for example, was that galaxies were smaller in the past, which was made by the simple method of noticing that the long-ago galaxies on these images were smaller than the ones we see around us today.

By the time we set off on our Grand Tour, there were fields like this dotted around the sky, most of them much larger than the minuscule area of sky covered by the Hubble Deep Field. There were long-exposure images of most of these fields in almost every waveband.

But not in the submm waveband. Besides the SCUBA image of the Hubble Deep Field, there were no images of any of these fields in the sub-mm waveband. The big discovery made from the first sub-mm images of the sky with SCUBA was of the 'sub-mm galaxies', galaxies that are so shrouded in dust that they are invisible even in *Hubble* images with ultra-long exposures (Chapter 4). Almost 30 years after the original discovery, it still seems certain that these galaxies, with stellar birth rates up to 1,000 times greater than in our galaxy today, are the ancestors of present-day elliptical galaxies. But in all the SCUBA images made by all the different teams, which together only covered a tiny area of sky, there were only a handful of these objects. One of the main promises of the Red Book was that the sub-mm images of large areas of sky possible with *Herschel* would discover thousands of them.

In our meetings around Europe, it was always obvious what we would do. The main purpose of the meetings was to discuss which fields we should observe – along with enjoying a lot of delicious food. We put in a proposal to Matt's committee asking for a big chunk of the available GT to obtain submm images of several of the main fields. Since it was one of the Red Book's main themes, the committee gave us most of what we wanted.

[⁵⁵] They had to be approved by the SPIRE Steering Group but the choices were Matt's.

The meetings of SAG2 were less convivial. The leaders of this SAG were Walter Gear, also a professor at Cardiff, and Sue Madden, a distinguished scientist who works at the Atomic Energy Commission in Paris, which by some quirk of history also contains an astronomy group. Sue, an American who now works in France, and Walter, from an Irish background who works in the UK, were not a natural combination, but they got on just about well enough to run our SAG.

What we should do with our GT also wasn't as obvious as it had been for SAG1. Matt had asked us to design a programme to observe individual galaxies.

There are definitely plenty of them.

Galaxies are the natural features of the universe. Like the mountains and rivers of the Earth, every galaxy is different, the product of its individual life history and environment. When we look out into the universe we see a galaxyscape of roughly two trillion galaxies[1,***] laid out in a network of intersecting galaxy archipelagos, which astronomers have named the 'cosmic web'. At the intersections are the universe's grand tourist sites: its rich clusters. A single cluster (Figure 6.1) may contain several hundred galaxies, although most galaxies are in more homely surroundings, by themselves or, like our own galaxy, in a small group.

FIGURE 6.1 *Hubble* picture of the cluster Abell 2218 (the faint arcs are the images of galaxies behind the cluster that have been stretched by its gravitational field – Chapter 11).

Credit: NASA, Andrew Fruchter and the ERO Team [Sylvia Baggett (STScI), Richard Hook (ST-ECF), Zoltan Levy (STScI)]

*** We can only see galaxies out to a horizon set by the finite speed of light and the age of the universe – beyond the horizon it would take longer than the age of the universe for the light to reach us. There are about two trillion galaxies within the horizon.

They vary wildly in size, from the giant elliptical galaxies at the centres of clusters (Figure 6.1), which contain over 100 billion stars[2] to some of the titchy dwarfs that circle our own galaxy containing only a few thousand.[3] There are several different types. The galaxies seen in an optical image are mostly massive galaxies, the spirals and ellipticals (Chapter 4), but most of the galaxies in the universe are dwarfs, which are less luminous and harder to see. Our own cosmic village, the Local Group, contains over 100 galaxies, but apart from the three large spirals[†††] all of them are dwarfs. Dwarf galaxies are mostly irregulars, a catch-all category that contains galaxies with no particular structure (Figure 6.2). Within each category, there is often intriguing variation. Edwin Hubble, the first astronomer to explore this world, noticed that the spiral galaxies range from ones with a large central stellar bulge and tightly wound spiral arms such as M81 (Figure 6.3) to ones like M83 (Figure 3.5) with a tiny central stellar bulge and loosely wound spiral arms.

We know, from observations in other wavebands, that spirals and irregulars contain large amounts of gas and dust, which explains why the stellar birth rate in these is much higher than in ellipticals (Chapter 4). We know it is gravity that keeps the stars from escaping from a galaxy, with the orbits of the stars in a spiral mostly being in a plane, like the planets around the Sun, while the stars in an elliptical whizz around in all directions like a swarm of bees. We know that most of the galaxies in the centre of a rich cluster like Abell 2218 (Figure 6.1) are ellipticals or lenticular galaxies, a type of galaxy with a huge bulge, hardly any disk and no spirals arms, such as the Sombrero galaxy in Figure 3.4. We know that most of the matter in a galaxy is not the normal stuff that makes up the stars and planets – protons and neutrons – but the dark matter whose composition is one of the big mysteries of modern astronomy.[‡‡‡] We know that the stars in the bulge of a spiral galaxy like M81 (Figure 6.3) are mostly old, red and with very low masses but that there are a lot of high-mass blue stars in the disk.

But this is all descriptive stuff: what a galaxy looks like, how big it is and how much gas it contains. A lot of the deeper things about galaxies we just don't understand.

We don't know, for example, why there are different types of galaxies. We don't even understand the most obvious thing about spiral galaxies: their spiral arms. Although there are several theories, we still don't know how these are formed. We don't know why most of the galaxies in clusters are ellipticals and lenticulars and there are very few spirals there. We still understand very little about the history of galaxies. I am pretty certain that the sub-mm galaxies are the children that will grow up to be present-day ellipticals, but that doesn't tell us everything – if I had only ever seen adults and babies I still wouldn't know much about human development.

The more we observe galaxies, the more questions there seem to be. Something that recently captivated astronomers everywhere was the first images of the black holes at the centres of galaxies. In a superb feat, not just of technical prowess but also of organisation,

---

[†††] Our own, Andromeda (M31) and the Triangulum Galaxy (M33).

[‡‡‡] We know that the dark matter is there because of the gravitational force it exerts on everything else. We know it can't be normal matter – protons and neutrons – because if the universe contained so much normal matter, the nuclear fusion reactions shortly after the big bang would have produced different amounts of the chemical elements from those we observe.

FIGURE 6.2   *Hubble* pictures of three irregular galaxies

Credit: NASA, ESA and the Hubble Heritage Team, A. Aloisi (STScI)

FIGURE 6.3 *Hubble* picture of M81.

Credit: NASA, ESA and the Hubble Heritage Team

an international team succeeded in linking together telescopes all over the Earth, creating the Event Horizon Telescope an interferometer (Chapter 2) capable of taking pictures with enough detail to see these central black holes. Figure 6.4 shows the first picture of the black hole at the centre of our own galaxy. Strictly, the picture is not of the black hole itself, because once light or any kind of radiation passes over the black hole's event horizon it can never escape. The picture is of the gas immediately outside the event horizon, but as the dark area in the centre delineates the event horizon, it's as close to being a picture as we can get of an object that does not emit any radiation.

But that just raises another question: How was this black hole created? The mass of our galaxy's black hole is roughly four million times the Sun's mass, but in some galaxies the central black hole is over one billion times as massive as the Sun. It is hard to think of a way of forming a black hole that is so large. There also seems to be a correlation between the mass of the central black hole and the mass of the surrounding galaxy. Why?

So there are a lot of questions about galaxies. In SAG2 we didn't think we would be able to answer any of the really big ones, but we hoped we might make a small contribution to answering some of them. We also knew we were in the fortunate position of having the opportunity to take the first-ever pictures of galaxies in a new waveband. The trouble was, we couldn't agree on which ones.

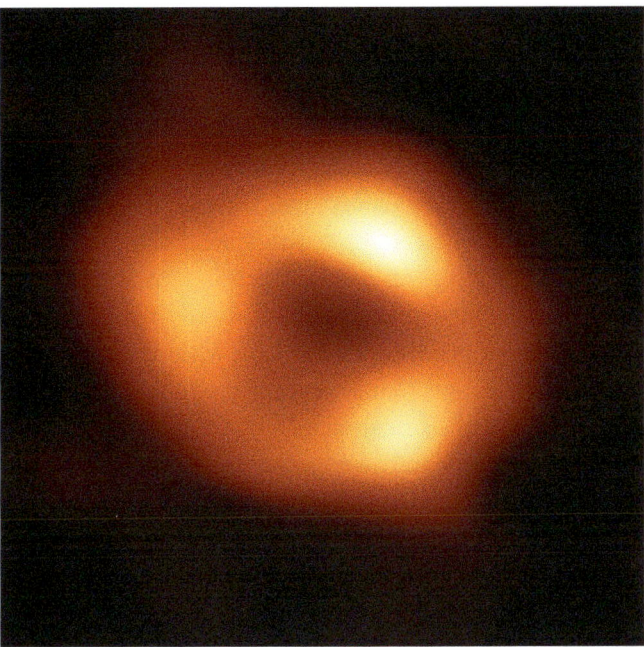

FIGURE 6.4 Image of the black hole at the centre of the Galaxy made with the Event Horizon Telescope. The darkness in the middle of the picture shows the region inside the event horizon.

Credit: EHT Collaboration

Everyone in the SAG had a different idea about how we should use the observing time. Sue Madden thought we should use most of it to observe dwarf galaxies. She argued these are special and worth observing because they contain an unusually large amount of dark matter relative to the normal stuff. For Alessandro Boselli, an Italian who also now works in France, it was the spiral galaxies – he argued we should observe these because they contain so much dust we would get some amazing pictures. I proposed that we simply choose a volume of space and observe all the biggest galaxies in it, which I thought would be a neat kind of cosmic census. Alessandro said that would be a mistake because most of these would be ellipticals, which contain very little dust, which would make them hard to detect. I thought that was the point. SPIRE would be so sensitive we would have the chance, for the first time, to find out how much dust there really is in ellipticals. Luigi Spinoglio, another Italian, wanted to observe a sample of galaxies with active nuclei – mini-quasars – and so on.

We took our arguments on a tour of beautiful European cities. After several meetings, we hadn't made much progress. I gradually realised there is no objective way to make a decision like this. Sue thought her arguments about dwarf galaxies were very strong. I thought a cosmic census was a great idea. But how do you weigh one against the other? It ends up being as much about emotion as anything else, which is why surprisingly often issues like this are settled by a vote.

In this case we didn't need to resort to a vote. After talking it through, we eventually agreed on a messy compromise. We merged Alessandro's and my idea. We did decide to do the cosmic census, observing all the biggest galaxies in a selected volume of space, but we also proposed to observe the smaller galaxies in this volume as long as they were spirals. We decided to call this survey, of 323 nearby galaxies, the *Herschel* Reference Survey. We also included in our proposal a project to observe a sample of dwarfs, which made Sue happy. And we included a third project, proposed by the Canadian astronomer Christine Wilson, to observe some of the very closest galaxies, a project we knew would produce sensational pictures. So after two years of meetings, we did have a plan.

But by this time I was finding it all a little boring. I hadn't really enjoyed science by committee (Matt hadn't mentioned that), but I eventually realised it was bigger than this. I thought the GT plan was too unambitious. *Herschel* would be the first telescope to explore the sub-mm waveband, a waveband that was almost virgin territory. Surely the point of a telescope like this is to look for the unexpected. All the GT projects were based on things we already knew (half the galaxies in the *Herschel* Reference Survey had been discovered by William Herschel, for heaven's sake). What I wanted to do was forget all about that, forget what we already knew: do what the first radio astronomers had done – survey a big area of sky just to see what's out there. I wanted to do the survey Loretta and I had talked about in my office ten years before.

The problem was that I had done some calculations. It just wasn't possible. Matt had built an instrument to survey a large area of sky, but not one large enough for the survey Loretta and I had discussed.

I lost contact with Loretta when she left Cardiff in 2004 to take a job as a lecturer at the University of Nottingham. During the next few years, while I was doing my tour of European committee rooms, Loretta's life was going through some big changes.

She quickly realised that the treadmill of a UK university job was not for her. Falling in love with New Zealand on a holiday, she decided she wanted to emigrate there and start a new outdoor life among that country's spectacular scenery. She applied for a place at a teacher-training college in New Zealand, which was the only way she could find of getting a visa. She was accepted, resigned from her job, sold her house, and bought a plane ticket, planning to move there at the end of the summer.

This was not good timing. During that summer she began to date another astronomer at Nottingham, Steve Maddox. She moved into Steve's house in Nottingham and turned down the place in the college. But now she had no job. She thought she had burnt all her astronomy bridges, but for something to do, she started going to seminars at the university again. One day she heard a seminar by the astronomer Jim Dunlop, a friend of ours who works at the University of Edinburgh. In the pub after the seminar Jim offered her a way back. He had some money left on a grant. He offered to pay her as a postdoc for a few months in Edinburgh to plan some observations he was hoping to make with *Herschel*.

In January 2007 out of the blue, Loretta phoned me from Edinburgh. She asked me if I was planning to go to the workshop to discuss open-time key projects which was being held at ESTEC the following month. I said, yes, of course, everyone who was interested in using *Herschel* would be going.

The open time, the OT, was the two-thirds of the *Herschel* observing time that had not been stitched up by the instrument teams. Anyone, anywhere in the world, could apply for time. A large part of the OT had been reserved for 'key projects', big projects requiring over 100 hours of observing time.§§§ Big projects mean big teams, and big teams have to get together somehow. The workshop was Göran's idea for how to do this. We would all go to ESTEC, hook up with astronomers interested in doing similar projects, get together in teams – speed-dating for astronomers.

Loretta asked me if I had looked at the sensitivity figures for SPIRE recently. She had needed to look them up to do some calculations for Jim's observing programme. There had been a big change. The sensitivity figures had ultimately come from Matt Griffin, and his estimates were based partly on the prediction by the engineers for the sub-mm emission from *Herschel's* primary mirror – the more sub-mm radiation from the mirror, the harder it is to detect the sub-mm radiation from a faint galaxy. Matt had been very conservative in his sensitivity estimates because, in his careful way, he didn't want to make a promise to the astronomers he couldn't deliver. Recently, the prediction for the sub-mm radiation from the mirror had changed. It had gone down by a lot. Matt had redone his calculations. SPIRE would now be much more sensitive than everyone had previously assumed. Loretta said that the survey of the large area of sky we had talked about during her PhD days was now possible.

A month later, along with 245 other astronomers, we found ourselves in the low office buildings in the middle of the flat Dutch countryside that are the home of ESTEC, the first time either of us had been there. The plan for the two-day workshop was that Göran and the PIs of the three *Herschel* instruments would talk first, describing the current state of the telescope, the latest sensitivity estimates, the observing modes of the instruments and so forth – everything an astronomer needs to plan an observing programme. Then the astronomers would talk. Göran had only given each astronomer a short time to talk, just long enough to sketch out the observing programme they were interested in doing. The schedule would allow plenty of time for informal discussion, meeting like-minded astronomers – the hooking-up bit.

Loretta was down on the schedule to give a talk on Jim's project, which no longer had any point because the new sensitivity figures meant that one of the GT observing programmes would now do the science that Jim had wanted to do. We agreed that at the end of her now redundant talk she would show a single slide proposing a very big survey of the sky and asking anyone interested to see us.

Göran, Matt and the other PIs gave long talks full of technical details. After I had drifted through these, trying to concentrate just enough to pick out the few facts that were important for the astronomers, it was our turn. One astronomer after another came to the front and described what they wanted to do with *Herschel*. Loretta gave her talk and showed her slide.

As we all spilled out of the conference room at the coffee break, even before we had found the coffee, Loretta and I found ourselves at the centre of a huddle of a dozen astronomers. They were all interested in the idea of the big survey. Gianfranco de Zotti, an Italian

---

§§§ Göran attributes the original idea for reserving some time for key projects to Reinhard Genzel when he was chair of the FIRST Science Advisory Group (Chapter 5).

astronomer, said the survey would be perfect for a project he had been thinking about. He wanted to look for gravitational lenses (Chapter 11), and he said that if we did this survey, we would find hundreds of these cosmic magnifying glasses.

I already knew Gianfranco a little from previous meetings, but this was the first time I had met the American astronomer, Asantha Cooray. Asantha is usually more interested in cosmology, the properties of the universe as a whole, than individual galaxies, but he said we had missed something important. Loretta and I were most interested in using the survey to see what was out there in the nearby universe, but Asantha said the new sensitivity figures were so good that, even with the short exposures we were planning, we would also find thousands of the high-redshift sub-mm galaxies discovered in the SCUBA surveys. This was the aim of the SAG1 team as well, but Asantha said the survey we were proposing would be better for finding the most luminous sub-mm galaxies – the ancestors of the biggest ellipticals in the universe today.

Everyone had their own idea of why a big survey was such a good idea. Dave Clements, who Loretta and I knew well because he had been a postdoc in Cardiff, was planning to work mostly on the data from *Planck*, which would be *Herschel's* travel buddy on the Ariane 5. *Planck* was primarily a cosmology mission, but it should also detect some of the same dust-enshrouded galaxies as *Herschel*. Its pictures, though, because it was half the size of *Herschel*, would show hardly any detail – the galaxies would appear as the blobbiest of blobs. Dave realised our big survey would give him more detailed pictures of the galaxies. Stephen Serjeant, another British astronomer, wanted to use the survey to look for rare objects. Others in the huddle just liked the idea of the first sub-mm survey of such a large area of sky, because we all knew it would be this survey that would find the new kinds of beast if they were out there.

Over the next two days, in the intervals between the talks and during the workshop banquet, we roughed out a plan. Once we realised how many things it would be possible to do with a survey of a large area of sky, we became worried that somebody not at the workshop might have the same idea. To avoid two competing teams, we fired off emails to everyone we knew describing the idea to those who were not at the ESTEC meeting and announcing that we would hold a workshop in Cardiff to discuss it further.

Then there was the question of how large we should go. The mission would only last three years and there was no way we could survey the whole sky. At the banquet, somebody – nobody can remember who – said, why don't we survey a thousand square degrees? That seemed about right. It would need 10% of all the time that had been set aside for the key projects, which was about as much as we thought we could get away with. It also gave us a catchy title: the Thousand-Degree Survey.

We also needed to decide who would lead the project, which is always awkward. It had been Loretta's and my idea, but Gianfranco argued that I should be the formal PI because Loretta, as a postdoc, was too junior. But Loretta and I agreed that even if I was the PI in the eyes of ESA, we would jointly lead the project.

During the workshop in Cardiff, at another workshop at Imperial College, and through hundreds of emails, we sorted out more details. A big question was which areas of sky, 'fields', to choose for the survey. Because the dust in our own galaxy emits a lot of sub-mm radiation, it was obvious that the fields should be well away from the Milky Way. It was also obvious that we should choose, if possible, ones that had already been surveyed in other wavebands – as the SAG1 team had done. Our problem was the size of our survey; there had been very few surveys, in any waveband, that had covered 1,000 square degrees.

We found a partial solution when Loretta, while on an observing run in Hawaii, bumped into another British astronomer, Simon Driver. Simon was leading a team that had started another big survey, and his team had gone through a similar discussion.

As astronomers, we navigate around the sky using a system very similar to that used on Earth. The right ascension and declination we use are almost exactly the same as longitude and latitude, although one difference for us is that east and west are reversed because we are viewing the celestial sphere from the inside. The celestial sphere, just like the Earth, has poles and an equator, which are just the projections of the Earth's poles and equator (Figure 6.5). Another important circle on the celestial sphere is the ecliptic,

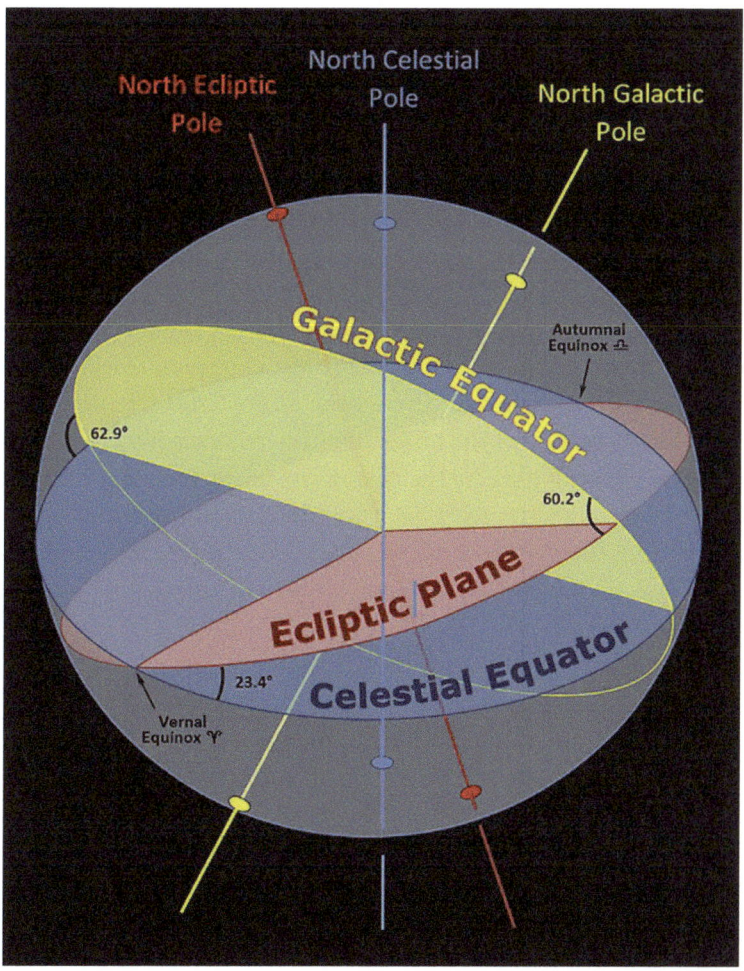

FIGURE 6.5   The celestial sphere.

Credit: Jim Slater

which is the projection of the plane in which the Earth and the other planets orbit the Sun (the ecliptic and the celestial equator are at an angle[⁵] because the Earth's axis is not perpendicular to this plane). The Galactic equator, which traces the disk of the Galaxy, is the third important circle and the Galactic poles are the points in the directions perpendicular to the disk.

Simon and his team were using a telescope in Australia to measure redshifts for hundreds of thousands of galaxies. Their galaxies had been selected from three fields on the celestial equator, which has the practical advantage that objects on the celestial equator can be observed with telescopes in both the northern and southern hemispheres. We realised these redshift measurements might be very useful for our survey and, since these fields were well away from the Milky Way and all its dust, we decided to use them for our survey as well. They covered 150 square degrees, which left us with the problem of which fields to use for the other 850 square degrees. There was no perfect choice, so we chose two large fields near the Galactic poles, which at least were well away from the Milky Way.

By the time we came to write the proposal, the team included 101 astronomers. The time allocation committee accepted our proposal but gave us only 600 hours rather than the 1,000 we had requested, because, they said, it wouldn't be possible to schedule the observations for such a large survey in the three-year lifetime of the mission. We reckoned that this would be still enough to cover almost 700 square degrees of sky, ten times larger than any other survey planned with *Herschel* – plenty enough sky to find any new beasts if they were out there. We were very happy.

In my pre-*Herschel* days, I usually had only a handful of collaborators in any project, so it never mattered too much if I messed things up; only a few people would care. But this was now becoming big science. By the end, there were almost 200 scientists in the team from 14 different countries. The value of the observing time was scarily large: if I multiplied our fraction of the total observing time by the cost of the telescope, 22 million euros.[⁎⁎⁎⁎] As the official PI, I also had to sign a letter to ESA, which stated that I would take personal responsibility for transforming the raw data from the telescope into data products – images and catalogues of the sources – and release these to the community no later than 18 months after the final observation.

We had more meetings to decide how to do this.

There were two instruments on *Herschel* that took pictures: SPIRE in the sub-mm waveband and PACS,[††††] which took pictures in the far-infrared waveband. We decided to observe our fields with both cameras, which meant we would have images of the sky at five different wavelengths.[‡‡‡‡] The downside of this decision was that we would need to reduce the data from two new instruments, which is always a challenge because with the data from any new instrument there are always quirks and problems that need to be sorted out. I volunteered my group to reduce the SPIRE data, which really meant my postdocs Robbie

---

[⁵] 23.4 degrees.
[⁎⁎⁎⁎] I doubt it would have fetched that much on eBay.
[††††] The Photodetector Array Camera and Spectrometer. There will be a test at the end of this book.
[‡‡‡‡] 250, 350 and 500 micrometres with SPIRE and 100 and 160 micrometres with PACS.

Auld and Simon Dye; Rob Ivison, a professor at the University of Edinburgh, volunteered to do the same for PACS; and Loretta and Steve Maddox volunteered to do the job nobody yet knew how to do: figure out which galaxies were emitting the sub-mm radiation.

Any big project needs rules. We wrote ourselves a constitution. The constitution stated that the project would be run by an executive: Loretta and me, the two PIs, and ten others. The executive would manage the process of transforming the raw data into usable products, organise the scientific projects we would do with the survey, adjudicate disputes between team members and so on. We designed the rules so that the people who got the first chance to write papers would be the people who did the work processing the data, which mostly meant the postdocs. We even gave ourselves the power to change the constitution if the rules didn't seem to be working.

FIGURE 6.6   Our logo

Credit: The *Herschel* ATLAS, Simon Dye

Since we would no longer be able to survey 1,000 square degrees, the project needed a new name. We sent out an email to the team asking for suggestions and then held a vote. We chose the *Herschel* Astrophysical Terrahertz Large Area Survey, very forgettable but with a catchy acronym: the *Herschel* ATLAS. My postdoc, Simon Dye, designed a logo for the survey (Figure 6.6).

We had a plan for reducing the data. We had a constitution. We had a name. We had a logo.

There was now just the matter of the launch.

CHAPTER 7

Minding the Heavens

I N THE LATE 1700s in Paris, in a time before electric light and *Netflix*, there was not much to do in the evening. But you could always go out to the observatory and watch the astronomers dance.

The dance started with one of the two dancers on top of a tower and the other on the ground about one hundred metres away. The one on the tower didn't do much apart from some occasional shouting. The one on the ground was more interesting. He started the dance by staring intently at something in the sky. Then after a minute or two he teetered on one leg, jumped from side to side, and then usually fell over. Then he got up and started again. There was a lot of swearing. It didn't look as if the dance was much fun, but it must have been entertaining to watch.

Its cause was a dead end that astronomers were trapped in for almost two centuries. At the time, astronomers were still using refracting telescopes, telescopes with lenses, which they had been using since the invention of the telescope in 1608. These still have their uses today, but one of their big problems in those early days was that a lens bends light of different wavelengths by different amounts, which had the effect of turning a crisp image of a star into a multi-coloured blob.* Astronomers were only able to avoid this problem by using lenses with almost no curvature, virtually flat pieces of glass, which bent the light by only a very small amount. But to get high magnification with such flat lenses they needed very long telescopes, which were hard to build and even harder to manoeuvre.

Then someone had the bright idea of doing away with the telescope's tube and making a telescope out of only lenses. In one of these tubeless telescopes – a deconstructed telescope – one lens was mounted on a tall building and the astronomer on the ground looked through a second lens to see the image formed by the first lens. It was exceptionally difficult to get the second lens in the correct position to see the image, and it was equally difficult to keep it in position – hence the astronomer's dance – but it was just about possible. A few big discoveries were made with these 'aerial telescopes', and the one at the Paris Observatory was used by Giovanni Cassini to discover two of the moons of Saturn.

---

* Eventually opticians discovered how to make lenses by combining glass with different refractive indices so that light of different colours came to the same focus.

DOI: 10.1201/9781003195290-7

But one big disadvantage was that while it was possible to magnify a single object, it was only possible to see a tiny area of sky around it. And they must also have been unbelievably difficult to use.

Eventually, after almost two centuries, one astronomer found a way of escaping from this technological cul-de-sac.

William Herschel was the man, the magus, the source. So many of the things we do in modern astronomy can be traced back to him. When Göran Pilbratt needed a name for the new telescope, his was the obvious one to use. He was the one who discovered infrared radiation and, as someone born in one European country but who worked in another, his name was nearly perfect for a pan-European telescope.

But he did so much more than this. Loretta and I wanted to survey a large area of sky to look for new types of objects. William Herschel was the first person to do a survey of this kind. He was also the first person to notice the places in the Milky Way where there are no stars – 'holes in the heavens' he called them – which we now know are caused by interstellar dust, one of the themes of this book. Two of its other themes are the origin of stars and the origin of galaxies – and Herschel was also the first person to try to understand these. He also transformed astronomy as a pursuit. His career contained a series of firsts: the realisation of the importance of big telescopes; the first catalogue of galaxies; the first survey; and even the first big grant application. In writing a book about the telescope, it is hard not to get enthralled by the man, he enters the story in so many places. In my interview with Göran Pilbratt, the *Herschel* Project Scientist, we started discussing the telescope but ended up gossiping about the astronomer as if he was one of our colleagues.

So much the same but also so many differences. Modern astronomy is a world of computers and screens. I could spend a whole observing run without even seeing the telescope, but Herschel looked through one. And without a camera, the only way he could keep a record of what he saw was by drawing a picture (which would have ruled me out as an astronomer then). He was the greatest observer of the nineteenth century and probably of all time. He seems to have had an amazing ability to adjust his eyes to the observing conditions, so he was able to see things right at the limit of human perception, without falling into the trap of seeing things that weren't there – as the observers later did who thought they saw canals on Mars. For a modern astronomer like me, he sets an impossible standard. And he made all his own telescopes which makes it even worse.

The thing I find most humbling is the sheer physical and mental endurance it must have taken. I may tell myself that astronomy is an adventurous pursuit, but my observing runs are short, I drive to the summit, I sit in a well-heated control room, and the most I have to complain about are jetlag and lack of sleep. But Herschel had to be there, night after night whenever the weather was clear, out in the open on a platform high above the ground, sometimes in subzero temperatures, doing everything himself. And during the day he was not able to relax like a modern astronomer. There was the telescope to maintain, new telescopes to build, and royal visitors to entertain to keep the money flowing. And he did this for almost 40 years.

I sometimes imagine William Herschel as a character in one of Jane Austen's novels, which is not as strange as it sounds because the setting for some of the novels is Bath and he was a music teacher there at the time, so he probably did give music lessons to young women like those in the books. Bath was the place where he became an astronomer....

He was born in Hanover, then one of the separate states in Germany. His father and his brothers were all musicians and made a living serving in military bands. Although Hanover was a poor place, its king also happened to be King of England, where there were more opportunities for musicians, and several members of the family made the voyage over, although all except William eventually became homesick and returned to Hanover.

The Herschels might have been poor but they were very clever. The males at least – his mother was illiterate and his youngest sister, Caroline, didn't receive much education because her mother wanted to keep her as the unpaid family servant – received a good, all-round German education, much better than the gruel of Latin and Greek that was taught in the schools in England at the time. If Caroline's memories are correct, an evening at home with the Herschels must have been an intimidating cultural experience. Apart from the music that filled the house, there were lengthy and excitable arguments about politics, philosophy and all the issues of the day. Caroline remembered hearing William, his oldest brother Jacob and their father discussing Newton, Leibnitz and Euler late into the night.

At the age of 18, William left this poor but cultured home forever. After some time in London, he based himself in the north of England, where he moved from town to town making a precarious living as a freelance music teacher, musician and organist. He became interested in astronomy, which was possibly a refuge from this hard-scrabble immigrant existence.

Everything that was known about astronomy and telescopes then was contained in only a handful of books, which meant that even a poor immigrant had a chance of making a contribution. It was also an exciting time because in less than a century, the human view of the universe had been completely transformed. Back in the seventeenth century, before Isaac Newton, the planets had been points of light, moving in a complicated dance around the sky. Now they were entire worlds – sometimes, in the case of Jupiter and Saturn, much bigger ones than the Earth. The stars themselves had been transformed from lights on a crystalline sphere to separate suns, which themselves probably had planets around them, strewn through a universe that had become unfathomably large. The planets and stars existed, but almost nothing was known about them. They were blank canvases. The universe was an empty space for the imagination.

As William in the late 1750s rode from gig to gig across northern England, he would often drop the reins and prop one of his astronomy books open on his saddle.[†] As the Sun

---

[†]   Two of the books William studied were *Astronomy* by the Scottish astronomer James Ferguson, which was first published in 1756, and *Opticks*, which was written by Robert Smith and first published in 1738.

or the Moon circled across the sky and his horse ambled across the moors, he began to read about astronomy and muse.[1]

Surely, if there were all those worlds out there, they must be teeming with life. Why else would the creator have made them? At this moment, there might be creatures on the Moon looking down at him – and if so the Earth would be their moon. What a different perspective that would give! Kant might be the last word in human philosophy, but what would the hypotheses be of a lunar philosopher or a philosopher on the Sun? The beautiful order of the universe was evidence for a creator, but what would happen to religion if we discovered there was life on these other worlds? Might all human religions be superseded by a new religion, a religion of the universe? There must be a multitude of discoveries waiting to be made. The universe was waiting for explorers….

We all need to make a living. William had to wait 16 years before he started out as an astronomer.[‡] In 1773 at the age of 35, when many modern scientists have already done their best work, he was hired as the organist of the fashionable new Octagon Chapel in Bath. He was soon the doyen of the Bath musical scene, then the premier one outside London. At various times he was also director of the orchestras at the Pump House and the New Assembly Rooms, the places where fashionable Bath society met to see and be seen. He was also a well-regarded composer, such a good one that some of his compositions are still available. He also had a house in New King Street, not the best address in fashionable Bath but still a perfectly respectable one.

Key to this new project were two people who were living with him there at the time. His brother Alexander had joined him in Bath because of its opportunities for musicians. Since his childhood, Alexander had been like William a tinkerer and builder of ingenious toys and devices, both brothers being natural craftsmen with none of the shame of the English middle classes at getting their hands dirty. The other person was their sister Caroline. William had recently succeeded in extracting her from the clutches of their mother by offering to pay for her replacement as the household drudge and by promising that he would train Caroline as an opera singer if she were allowed to come over to Bath. It is also possible he realised that he would have more time to spend on this new project, which would be on top of all his musical commitments, if Caroline took over the household duties in New King Street.

He started the project in the usual way at the time by buying some lenses and making his own telescope, but he soon encountered the old problem with refracting telescopes. In search of the greatest possible magnification, he made a telescope with a tube of pasteboard 20 feet long. This proved to be completely unmanageable, it bent like an elephant's trunk, and he was the only one ever able to see anything through it.

Then he heard about somebody in Bath who had made a different kind of telescope. Back in the seventeenth century, Isaac Newton invented the reflecting telescope, in which the light is brought to a focus by a curved mirror, although a lens in an eyepiece is still used

---

‡ Back in the eighteenth century, there were only a few people in Britain who made a living as an astronomer: the Astronomer Royal, possibly a couple of people at the Royal Greenwich Observatory, and that's probably all of them. Back then, if you wanted to be an astronomer, you first had to find a way of supporting yourself while you did it.

to view the image made by the mirror. The advantage of a reflecting telescope is that the light of all wavelengths comes to the same focus, so there is not the problem of a star being transformed into a multi-coloured blob.

In today's world, I rarely think about the mirror of my telescope. Nowadays the mirrors are made by big industrial companies, which use machines to grind a glass disk into the correct shape and then deposit a thin layer of silver onto the glass to make a reflective coating. If I think about it at all, it is when something goes wrong as it did with *Hubble* (Chapter 5). Back in the eighteenth century, though, any astronomer wanting to use a reflecting telescope first had to make their own mirror. The mirrors were made from speculum, an alloy of copper and tin, which is nowhere near as reflective as the silver used to coat the surface of a modern mirror. The astronomer had to grind a disk of speculum into precisely the right shape, and back in the eighteenth century there were no machines to do this. It is not surprising that most astronomers at the time preferred to use a refracting telescope.

The person William had heard about, a Quaker,[2] had taken up astronomy as a hobby but had found it too difficult and decided to sell his equipment. William made an appointment to meet him after chapel on Sunday 22 September 1773. After receiving some basic tuition on how to grind and polish a mirror, William left with all the tools and some half-finished mirror disks.

The house in New King Street now became a telescope factory. In one of the bedrooms, Alexander was using a lathe to grind disks of glass to make the lenses for the eyepiece; in the drawing room a cabinet maker was making the tube and stand for the telescope; and down in the basement, William had started to make the mirror, using a mould of horse dung, as one did at the time, to cast the speculum disk. All this was happening – they still had to make a living – among regular music rehearsals and lessons. With the tools, musical instruments, workmen and students traipsing through the house and the smell of horse dung drifting up from the basement, it cannot have been much like the sanitised Georgian interiors seen in the BBC adaptations of Jane Austin's novels.

It was William himself who made the mirrors for all his telescopes except for his final one. While in the house in New King Street, he cast the speculum disks in a furnace in the basement, which can't have been the safest thing in the world. The furnace did eventually explode, spilling liquid metal over the floor and splitting the flagstones.[§] The most challenging stage was grinding and polishing the disk so that the mirror had exactly the right shape. He found he had natural talent for the craft, and he invented several 'contrivances' for checking the shape of the disk's surface as he worked on it. The rest was just perseverance. He preferred to complete a mirror in one sitting because it allowed him to feel the mirror taking shape under his hands, and he was prepared to sit grinding and polishing a mirror for well over ten hours at a time. To endure the boredom, he made Caroline read him novels as he worked, and during one marathon 16-hour mirror-grinding session she put bits of food and drink into his mouth so he didn't need to take his hands away from the mirror.

---

§   The house is now the Herschel Museum of Astronomy. It is still possible to see the cracks in the flagstones.

From the moment of his first observation, William had supreme confidence in the quality of his telescopes and in his ability, despite no formal training, to interpret his observations. One of the first targets he noted down in his brand-new observing book was the Orion Nebula, the smudge of light visible below Orion's belt. There was a drawing of the nebula in one of the books he had used to learn about astronomy⁑. Looking at the nebula through the telescope, William thought it looked different from the picture in the book and decided the nebula must have changed in the time since it had been drawn. He was wrong about this, but it brought into his mind the idea that the objects in the sky, like those on earth, are not eternal, that they change – the idea of evolution.

Another of his early targets was the Moon, where he hoped to see the life that he was sure existed there. In one of the dark areas in the Moon, the Mare Humorum, he believed he could see forests or at least 'large growing substances'. Although there was no evidence of people, he was convinced that the thousands of small craters must be the cities of the Lunarians.[3]

The news of the amazing telescope gradually spread. A doctor in Bath, William Watson, happened to see William while he was observing and asked to look through the telescope. Enthralled by what he saw, he spent the whole night talking to William and became a frequent visitor to New King Street. Through Watson, who was a member of the Royal Society, Neil Maskelyne, the Astronomer Royal, heard about the telescope and decided to come down to Bath to see it for himself. Watson briefed William carefully on how to deal with Maskelyne, a difficult character, and especially not to mention any Lunarians. William cheerfully ignored his advice and when the two met Caroline heard through the wall what seemed like a violent argument, but when Maskelyne looked through William's telescope, he saw that many bright stars, including the Pole Star itself, were double stars, something he had never been able to see with his own telescopes. Who cared if the crazy German thought there were people on the Moon.[4]

One advantage of his new telescope was something I suspect William himself did not realise immediately: the telescope's field-of-view was much larger than for any other telescope of the time.⁑⁑ This was important. It raised the possibility of surveys – of finding new types of objects.

In August 1779, five years after he had opened his first observing book, he started a project to observe all 6,000 stars in the sky, with the not particularly exciting aim of finding out how many of them are actually double stars.[5] Eighteen months later, on 13 March 1781, he noticed an interesting object.

The object had a fuzzy appearance and over the next few nights he discovered that it was moving relative to the stars. From its movement, he realised it must be in the solar system. He first thought it might be a comet but eventually decided that its outline was too sharp. There doesn't seem to have been a Eureka moment,[6] but it gradually became clear to

---

⁑   *Opticks* by Robert Smith, published in 1738
⁑⁑ The reason is the f-number, the ratio of the focal length of a telescope to the diameter of its mirror. All Herschel's telescopes had small f-numbers, which gives a large field-of-view. Most refracting telescopes at the time, especially the aerial telescopes, had much larger f-numbers.

everyone that it was a new planet. On that Tuesday, from his back garden in Bath, he had discovered the first planet to be discovered in modern times – Uranus.

The discovery of Uranus set astronomy off in a new direction. From that moment, William spent much less time observing individual objects and most of his nights on systematic surveys of the sky, writing down everything he found. Nowadays, a survey is usually the first thing that is done with a new telescope or instrument. Carry out a survey. Look for new objects.

The discovery of Uranus also changed his life in a practical way. Lobbied by William's new friends at the Royal Society, King George agreed to give him an annual pension as the 'King's Astronomer'. Leaving music behind, he became a professional. He and Caroline moved to Datchett (Caroline doesn't seem to have been consulted) and then to Slough, both not by chance close to the royal court at Windsor. Alexander stayed in Bath to work as a musician but often came down to Datchett to help with the telescopes.

For the next 40 years, the observatory built by the Herschels was the centre of world astronomy.[††] With the security of the King's money, William was able to start a programme of technological development that has continued until this day. He was the first to realise the importance of the amount of light gathered by a telescope, which sets the limit on how faint a star or galaxy you can see – how far out in space you can look. The primary mirror acts like a big bucket for collecting light photons, and so the sensitivity of a telescope ultimately depends on the area of its mirror. Nowadays, we can also increase the amount of light collected from a faint star or galaxy by increasing the exposure time on the camera, but in the eighteenth century there were no cameras. The size of the mirror was everything, and William was the first to realise this.

Astronomy's first big telescope, the one built by the Herschels in Datchett and used for their surveys of the sky, is shown in the drawing (Figure 7.1). The diameter of its mirror was 45 centimetres, which is roughly 100 times less than for the mirror of the Extremely Large Telescope, which as I write is under construction in Chile. But by the standards of the eighteenth century, the Herschels' telescope was a technological marvel. Apart from the unprecedented size of the mirror, made as usual by William himself, the telescope and its huge framework, made entirely out of wood, was almost as high as the nearby house shown in the drawing in which the Herschels were living. The telescope's size, in an era before motors, meant that it was extremely hard to move, and in practice it was usually kept pointing towards the south and mostly only moved upwards or downwards, like the barrel of a gun, which was done by winding a windlass at the bottom of the tube. The observer, to see through the eyepiece, needed to climb onto the observing platform high above the ground.[‡‡]

[††] The poet John Betjeman wrote, 'Come friendly bombs and fall on Slough' and Ricky Gervais set the *Office* there. Slough does not have the best reputation, but – note to the Slough tourist board – Slough was the Mauna Kea of the eighteenth century.

[‡‡] The telescope had a Newtonian design in which light is reflected off the primary mirror at the bottom of the tube, then reflected by a small angled flat mirror at the top of the tube into the eyepiece.

FIGURE 7.1   Astronomy's first big telescope. The Herschel house in Datchet is behind it. Caroline sat in a room in the house, possibly behind the ground-floor window in the picture.

This meant a single observer would have found it very hard to operate the telescope because of the constant need to clamber down the ladder to change the telescope's direction – almost guaranteed to lead to a serious accident, especially on an icy winter night.[§§]

The Herschels overcame this problem by working as a team.[¶¶] William stayed up on the observing platform, away from all light so that he could keep his eyes adapted to the dark. Caroline sat in a nearby room, which was possibly the room behind the ground-floor window in the drawing. On her desk, she had Flamsteed's Atlas of Astronomy, which contained accurate positions of all the bright stars, and paper and pen. She also had an accurate clock, constructed by Alexander, and an ingenious instrument that allowed her to monitor the motion of the telescope. This consisted of a cable connected to the telescope, which passed through the window and over a measuring grid on the wall of her room. If the telescope

---

[§§]   Health-and-safety officers would never have approved any of the Herschels' telescopes. Both William and Caroline did have serious accidents.

[¶¶]   This has only recently become clear because, in the convention of the time, all the scientific papers describing their results were single-author papers written by William himself (in these papers he occasionally refers to an anonymous assistant). But it is clear from visitors' accounts of their observing method[10] how much the big survey that changed our conception of the universe was a team effort.

moved upwards (northwards in the sky) or downwards (southwards), a pointer attached to the cable moved over the grid.[11]

When William saw an interesting object, he pulled on another cord, which alerted Caroline and she opened her window. William shouted down everything she needed to calculate the object's position and anything else worth writing down. If he had discovered a new nebula, for example, she was able to calculate its position using the stars in Flamsteed's catalogue. If a telescope is kept fixed in position, objects drift from east to west through its field-of-view, and so Caroline was able to calculate the nebula's east-west position from the difference in the times it and one of Flamsteed's stars passed through the field-of-view. She was able to calculate its position in the north-south direction by measuring the difference between its position and one of Flamsteed's stars using the pointer's motion over the grid in her office. After she had written everything down, she shouted it back to William to check she had heard him correctly. She shut the window and the observations resumed.[7,11]

The Herschels carried out their survey by making use of the drift of the sky through the telescope's field-of-view. They also made use of the muscles of another member of the team, a workman who spent the night cranking on the windlass so the telescope moved at a constant pace alternately upward and downward (Alexander devised a system that would ring a bell when he needed to change directions). During the course of a night, using the drift of the sky and the workman's muscles, they surveyed a strip of sky two degrees wide.[11] In 1784 and 1785, they covered most of the sky visible from Datchett in these two-degree strips – 'sweeps' as William called them.

A survey doesn't finish with the observations. A vital part of any survey is the conversion of the raw observations into a catalogue: the positions and other properties of the objects discovered in the survey. This was made harder in the Herschels' surveys by their ingenious method for measuring the positions of the objects.[8,11] To calculate the positions, Caroline needed to know the relationship between the inches moved by the pointer across the grid and the motion of the telescope. They gradually improved their knowledge of this relationship, finding that it changed depending on where the telescope was pointing in the sky. The trouble was that every time they improved their knowledge of this relationship, all the old positions needed to be recalculated, sometimes for over 1,000 objects. In a time before calculators, all this had to be done with pen and paper. Caroline did all this work during the day. 'Minding the heavens', she called it.[9,11]

One of their big discoveries was the size and shape of the huge assembly of stars in which we live. When he looked at the Milky Way, William would see its light dissolve into hundreds of stars within the telescope's field-of-view.[12] But when he looked away from the Milky Way, he would usually see only a handful of stars. This suggested to him a way of mapping this stellar assembly. If the density of stars in this assembly is roughly the same everywhere, the number of stars we see in any direction should be in proportion to the distance to the boundary in that direction. If the distance to the boundary in one direction is 100 light-years and in another is 10 light-years , we should see roughly ten times more stars in the former direction.[13]

Counting stars sounds easier than it must have been. If there were hundreds of stars in the telescope's field-of-view, it would have been impossible to count them all before some of them had drifted out of sight. What William did in these cases was count the stars in half of the field he could see through his telescope or, in the most crowded fields, one quarter, and then multiply the answer by two or four, respectively. During the survey he and Caroline carried out during 1783 and 1784, he counted the number of stars in 683 different directions. The greatest number he found was 588, the smallest number zero. Using these star counts, the Herschels were able to draw the first map of the stellar assembly, which we now call the Galaxy (Figure 7.2). They were also able to estimate its rough size: 15,000 million million miles or, in astronomers' units, about 3,000 light-years.[14]

I have called it a map, which is the way it is often described, but the historian of science Dr. Wolfgang Steinicke has recently shown it is really a cross-section of the Galaxy.[15] The 683 fields in which William counted stars were distributed all over the sky, so in principle he could have used these counts, each of which gave him the distance to the Galaxy's boundary in one direction, to create a 3D model of it. But this was the eighteenth century – no computers, no calculators, nothing but pen and paper and the human brain. Back then, the best the Herschels could do was choose a circle around the sky and use the star counts along this circle to draw a diagram like that in Figure 7.2, which is therefore really a cross-section.

Figure 7.3 shows how their cross-section, which was drawn from the counts in 127 separate fields around a circle, relates to the structure of the Galaxy as we understand it today. Their cross-section didn't reveal the full radial extent of the Galaxy's disk because they chose a circle that happens to lie an angle to the disk, although they did get the disk's thickness about right. But even if they had chosen a circle that lay along the disk, they would have drastically underestimated the size of the Galaxy because in these directions most of the stars are hidden by dust – yet another connection with the eponymous telescope.

The survey that the Herschels carried out during 1783 and 1784 changed the human conception of the universe.

And then the universe became even larger.

It was because of Caroline that William became interested in the nebulae. After they moved to Datchett, she had fewer household duties, so he made her a small telescope

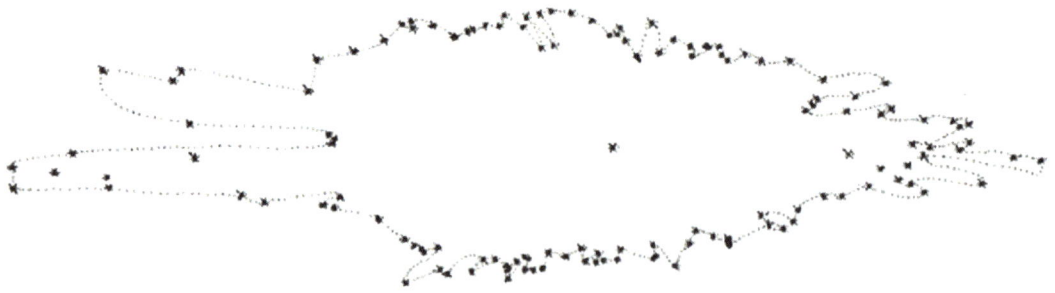

FIGURE 7.2   The Herschels' map (really a cross-section) of the Galaxy.

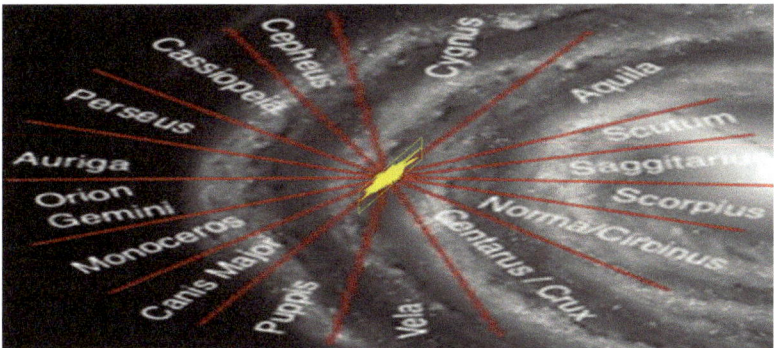

FIGURE 7.3 The Herschels' cross-section of the Galaxy in yellow superimposed on a modern representation of its disk. The Herschels missed the huge extent of the Galaxy because of the angle of their cross-section (although they would have missed it anyway because of the effect of interstellar dust), but their estimate of its size is not too far off current estimates of the thickness of the Galactic disk.

Credit: Wolfgang Steinicke

and instructed her to go out and look for comets. She became one of the most successful comet-hunters of all time, finding by the end of her life eight new comets.

For comet-hunters, the nebulae are always a problem – faint, fuzzy patches of light that look quite like comets except they don't move through the sky. A generation before the Herschels, another famous comet-hunter, the French astronomer, Charles Messier, had become so irritated that he made a list of 103 nebulae, so he would stop mistaking them for comets. When Caroline started searching for comets, she quickly found two nebulae that were not on Messier's list. William became interested.[16]

Nebula hunting was more fun than counting stars. While still counting stars with the big telescope, he also began to look for nebulae. All stars look the same except for the colours of some of the brightest ones,*** but every nebula he found was different. There were irregular nebulae like the Great Nebula in Orion but there were also nebulae with distinctiive shapes. There was one nebula that looked like a fan, another with the shape of a brush, and one that had a bright nucleus and looked like a comet. There were nebulae with a mottled appearance and nebulae that looked like 'cloudy stars surrounded with a nebulous atmosphere'. There were bright nebulae that seemed to have a faint one attached. There were even triple nebulae. Every nebula was different.[17]

He discovered that the best nights for nebula-hunting were ones in spring, especially nights on which there was no moon. As a pampered modern astronomer, I find it impossible to imagine fully what one of these nights would have been like but[18]…

*** The cells in the eye that register colours only do so at high light levels, which is why only the brightest stars seem to have colours.

Sometime in spring 1784, after an early dinner, he and Caroline got ready for the night's observations. Caroline went to her room. He climbed onto the observing platform. It was going to be a cold night, so he had put on three layers of clothing, and now he daubed his hands with onion juice to stop them from freezing when he handled the telescope. The sky looked clear, which was promising, and there was no moon. They had hit a rich stratum of nebulae the previous night and he felt very hopeful, although you never knew on cold nights like this because the seeing wasn't always very good.

They had finished a sweep the previous night, so now he shouted to the workman to come out and raise the telescope. The sky began to darken. He sat down in his observing chair and looked through the eyepiece. He was only able to see a few stars, slowly drifting across his view. As the sky got darker and his eyes adapted, more stars became visible. When it was completely dark, his eyes fully tuned, he began counting stars, yanking on the cord after each field of stars to alert Caroline and shouting down the results. In the first few hours he didn't see any nebulae.

But it was always difficult, finding nebulae. Ignorant people thought everyone saw in the same way, but he knew differently. The eye was an imperfect instrument and needed to be trained. He knew all the tricks of the eye, the distortion if he stared at a star too long, the importance of resting the eyes, of having a tranquil eye. There was an art to seeing. He was sure he could now see much more detail than when he first looked through a telescope. He had trained his eyes to see, really see, but it was still difficult finding nebulae, they were so faint.[19]

By ten o'clock he still hadn't found any, but he continued with his star counts, shouting down the results to Caroline.

And then, an hour before midnight, he discovered his first nebula. And then a few minutes later he discovered another, and then another. In 36 minutes, he discovered 31 new nebulae. It was the richest nebulous bed he had ever found. Occasionally, when he saw an interesting one, he drew a sketch. All the nebulae looked different, although drawing accurate sketches was so difficult – they were so hard to see.

But what were they? After much contemplation, he had concluded they must be like the Milky Way: huge assemblies of stars. When viewed through his telescope, the Milky Way dissolved into a multitude of stars, and since the nebulae were also faint patches of diffuse light, they must, by analogous reasoning, also be stellar systems. He was also sure he could see stars in some of the brightest nebulae, including the great nebula in Andromeda. He could not see the stars in the other nebulae, but he thought that this must be because they were too far away.[20] The distance to the edge of the Galaxy was almost incomprehensible, but the distances to the nebulae must be much greater. He reckoned that the faintest nebulae must be about 750,000 light-years away. That was so far away that the stars in the nebula might not exist anymore. The rise and fall of empires, the whole of human history, had happened in the time since the light left the nebula.[21]

It was now after midnight, the darkest and quietest time of the night. If he turned his head, he could see the light in Caroline's window. His eyes were tranquil, perfectly adjusted for discovering faint nebulae. It was getting cold. Sitting here now, the wood cold to his touch, he was finding nebulae never seen before by the eyes of men.

He was certain now he wouldn't see any changes in these huge stellar systems in his own lifetime. But these were objects in God's universe. They must be born, grow and decay like everything else. The nebulous beds, with their variety of forms, looked remarkably like the beds in a luxuriant garden, containing plants at every stage in their development. It might be possible to use the specimens he found to reconstruct the life history of a nebula. There must be a natural history of nebulae as there is of plants.[22]

His analogical reasoning seemed sound. These were all conjectures, of course, but bold conjectures are always important.[23] But how was he to prove them? Everyone accepted that the Milky Way is the combined light of myriad stars. The obvious way to prove that the nebulae, too, are vast stellar systems was to show that they too dissolve into stars when viewed through a big enough telescope. He was sure that he had already seen stars in the great nebula in Andromeda, and there were other nebulae with a mottled appearance that seemed to him on the verge of dissolving into stars.

If only he had a big enough telescope.

To prove to the world that the nebulae are vast assemblies of stars – galaxies like our own, although the word didn't exist then – the Herschels built an even bigger telescope. The new telescope would have a mirror 48 inches in diameter, which was much too large for a single person to make; grinding and polishing the new mirror would require machinery and a team of workmen. The project was also well beyond the Herschels' own financial resources. William wrote the first astronomy grant application, asking the king for the money to build the telescope. George III graciously granted him 2,000 pounds, enough to build the telescope and provide four years of running costs.

Every aspect of the project was challenging. As the great telescope gradually took shape, it became a favourite tourist destination for visitors from London and the royal court at Windsor. The weight of the mirror, about 1,000 kilogrammes, was almost the same as that of a small car. The mirror was cast in London, shipped upriver to Windsor, and then transported overland to Slough, where it was ground and polished by a team of 24 workmen. The telescope's tube, an iron cylinder 40 feet long with a diameter of 5 feet, was big enough for Caroline to walk along without stooping. About the size of a six-story apartment block, the telescope's wooden framework dominated the skyline around Slough. This was Big Science – the biggest astronomy project for half a century until the Earl of Rosse built an even larger telescope.

William had written the first grant application. He was also responsible for the first budget overrun. He had drastically underestimated the cost of the project and had to ask the King for more money. His majesty was not amused, but, after confrontations so mortifying they were concealed by the respectable middle-class Herschels for over a century, the King eventually relented and granted an additional 1,000 pounds to finish the project. Construction continued. More months were spent polishing the mirror. The visitors came and went. The huge iron tube was a favourite tourist stop, and in 1787 a banquet to celebrate the project ended with a musical crocodile dancing through the tube as it lay on the ground.

Two years later, in August 1789, a month after the fall of the Bastille in Paris, the telescope was ready for its first observations (Figure 7.4). There was a lot riding on its success. The King had given a large sum of money for its construction (twice) and the President of the Royal Society, Sir Joseph Banks, had spent a lot of his personal credit with the King in persuading him to support the project. As the great telescope rose above the Slough skyline, astronomers all over Europe wrote letters to William anticipating the observations that would be made with it. The telescope was already the wonder of the age, with its 'mighty bewilderment of slanted masts, spars and ladders and ropes, from the midst of which a vast tube, looking as if it might be a piece of ordnance…[lifted] its mighty muzzle defiantly to the sky'.[24] But it was also expected to deliver wonderful new discoveries.

FIGURE 7.4 The Herschels' 40-foot telescope. Even compared to telescopes today, it looks big.

The first night William used the telescope, 28 August 1789, everything went splendidly. He discovered a sixth moon around Saturn. A month later he discovered a seventh moon. When Sir Joseph Banks heard the news, he wrote a letter to William about the stream of wonderful discoveries that was sure to come from the telescope.

But this was its last one. The problem was the mirror. It took so long to warm up in the morning that it was prone to condensation, which tarnished it. Tarnishing had always been a problem with his mirrors, and it had been necessary to re-polish all of them now and then. But this new mirror weighed almost a tonne. It needed to be hoisted out of the tube by a crane, taken on a carriage to the polisher, polished for several days by workmen specially hired for the purpose, and then transported back to the telescope and hoisted back into the tube. It was also a risky business; when a wooden beam broke during the mirror polishing of 1807, Alexander and William were almost crushed to death. The new telescope was even less manoeuvrable than his other telescopes, making it less suitable for surveys. After the first month of observations, William himself almost never used it, preferring to use his smaller telescopes.

He was also disappointed when he looked at the nebulae. He had hoped that through the new telescope the nebulae would dissolve into individual stars, proving his theory that they, like the Milky Way, are vast stellar assemblies. It had been the whole point of building it. But through the new telescope, the nebulae remained obdurately as they always had been – faint patches of light. The new telescope was no better than the old one.[†††]

Then, only a year after his first observations with the new telescope, he discovered a nebula that collapsed the whole theory.[25] He discovered, with his old 22-foot telescope, a faint nebula around an individual star. It was clear this nebula was not the combined light from many stars but something else. Never afraid to change his mind, he made a bold new conjecture: that the nebulosity must be light from a fluid that was gradually coalescing to form the star – thus simultaneously asking, for the first time, the question of how a star is born and making an attempt to answer it. But it also meant the end of his grand theory of the universe.

With the end of his theory, William went on, as we all do, to other projects. He became interested in the Sun. After a series of careful observations of sunspots, he became convinced that the sunspots must be gaps in the Sun's atmosphere, revealing the solid surface below. Using the style of analogical reasoning he used in all his scientific work, he concluded there must be people living on its surface.[26] To observe the Sun safely, he used filters that blocked out most of its light. He noticed that he could still feel the Sun's heat through the filters, with the heat greatest when he used a red filter. To investigate the reason, he constructed an elegant experiment in which he passed sunlight through a prism, using a thermometer to measure the temperature increase produced by light of different colours (Chapter 2). He found that the temperature rose even if the thermometer was held beyond

---

[†††] It is easy to show with modern hindsight that he would have needed a much bigger telescope to detect stars in any galaxy further away than Andromeda, the nearest big galaxy. The mottled appearance of some of the nebulae that he had noticed in his earlier observations was probably created by star clusters rather than individual stars.

the red end of the visible spectrum, discovering infrared radiation and showing that major discoveries can come from the dodgiest of beginnings.[27]

But all this time the big telescope was still there, towering above the streets of Slough. It was impossible to simply forget about it. All the tourists still wanted to see the wonder of the age. For almost three decades, there was a stream of royal visitors from the nearby court wanting to look through the famous telescope. For three decades, the Herschels had to think of reasons why it was not possible that evening (the mirror was tarnished, the mirror was being polished….) and fob them off with a view through one of the other telescopes. It was all rather embarrassing. The telescope mouldered above Slough for almost 50 years. Eventually, the woodwork deteriorated so much that the structure became dangerous and William's son, John, also an astronomer, made the sensible decision to lower the tube to the ground.

There is a moral to this story. Even if the scientific reasons for building a telescope are very strong; even if the team that built the telescope is full of exceptionally talented people; even if all the technical challenges seem to have been overcome – all of which were true for the telescope we were now waiting to use – until the first observations nobody knows whether it will work or not. Until first light, nobody knows anything.[28]

CHAPTER 8

First Light

CAROLE TUCKER WAS IN a viewing cabin seven kilometres away from the launch site. The ESA management was only willing to pay for a few important people from the *Herschel* team to be there at launch, and Carole, a junior member of the team, was not one of them. The only reason she was there was because she was also in the *Planck* team, which had somehow managed to take everyone along. The important *Herschel* people – Göran Pilbratt, Matt Griffin and the other two instrument PIs – were in the VIP viewing cabin only five kilometres from the Ariane 5.

Matt Griffin wasn't worried. He was an engineer and trusted the engineers who had built the Ariane 5. There had now been 29 launches in a row without a hitch. What he did worry about was whether the instrument he and his team had built, which had cost millions of pounds of taxpayers' money, would work. At least, if the Ariane 5 exploded on launch, nobody would ever know.

As the rocket lumbered into the air, Göran Pilbratt was momentarily startled. There was none of the noise you hear on TV. Then he realised: light travels faster than sound. Fourteen seconds later, with the rocket already high in the sky, the rolling thunder of the launch reached him across the jungle. A couple of seconds later the rocket vanished into the clouds. After a few seconds, it emerged, now travelling at a slight angle on a single pillar of flame. The spectacle for him did not really measure up to all that work and emotion. In the non-VIP viewing cabin, the crowd around Carole erupted. The spectacle of a crowd of middle-aged scientists and engineers whooping and hollering and madly hugging each other was strangely more affecting than the point of light now streaking across the sky.

After the party, everyone went home. Back at ESTEC, Göran Pilbratt was now working seven days a week. At eleven every morning, he was there in the ESTEC video-conferencing suite for a conference call with the three instrument teams and Mission Operations in Darmstadt to get ready for the next contact period. The ESA ground stations were only in contact with the spacecraft once a day and everything had to be done during this three-hour period.[*]

---

[*] At the time, there were two large ESA ground stations, the ones used to communicate with distant spacecraft, in Spain and Western Australia. ESA had only allotted Göran and his team one three-hour window each day to talk to *Herschel*, which awkwardly – there are a lot of spacecraft – happened at a different time each day.

DOI: 10.1201/9781003195290-8

Every time they made contact, they never knew what they would find. If everything was nominal – OK in space talk – they uploaded the instructions for the next 21 hours, during which the spacecraft would be operating remotely. If they found anything wrong, they would also need to fix it, either themselves or by waking up some expert in one of the *Herschel* teams around the world – all in three hours.

Herschel was now on its way. Its destination was well beyond *Hubble*, which is in such a low orbit it is hardly above the atmosphere. It was heading for an orbit around the second Lagrangian point, L2, one of the five Lagrangian points at which a small object can orbit the Sun while remaining in a fixed pattern with both the Earth and the Sun. L2 is four times further away than the Moon – much too far to send astronauts up to fix things if anything goes wrong.

Launch Day plus 2: The spacecraft crossed the orbit of the Moon, 384,000 kilometres from Earth.

Launch Day plus 4: The engineers turned on the telescope's heating system. As the telescope got colder, the heating system would stop condensation on its mirrors.

Launch Day plus 5: An Argentinian amateur astronomer, Gustavo Muler, took a picture of *Herschel* and *Planck* (Figure 8.1), the last ever picture of the two spacecraft.

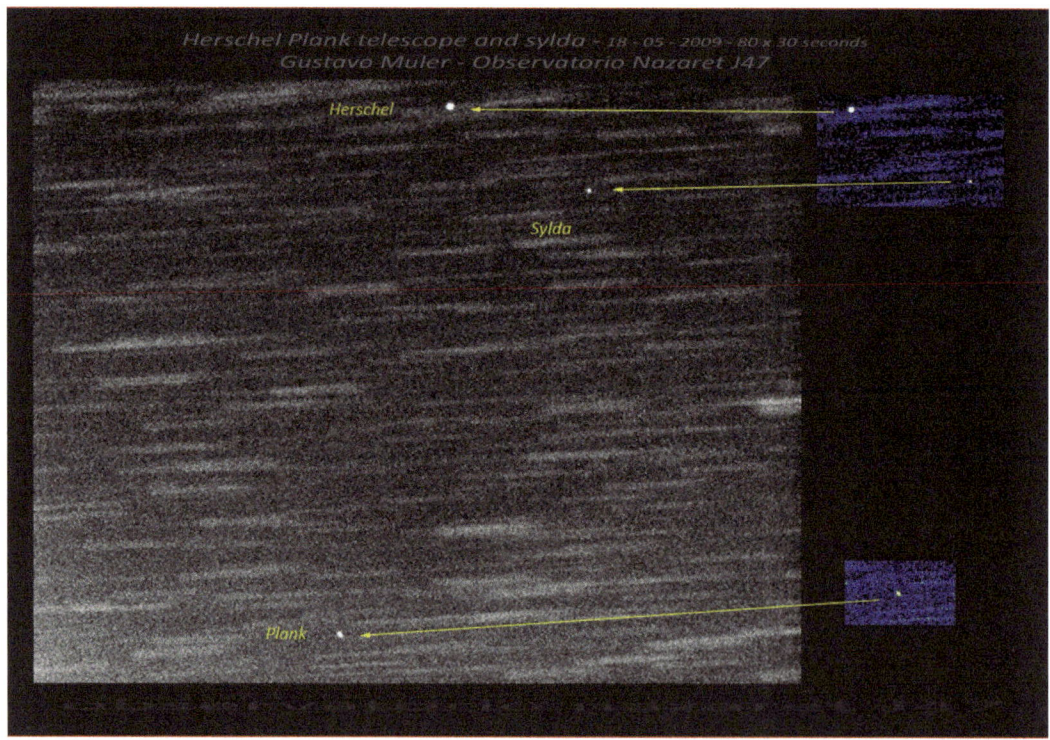

FIGURE 8.1 The last ever picture of *Herschel* and *Planck*. The third dot is the Sylda, the metal cage in which the two spacecraft rode into space.

Credit: ESA, Gustavo Muler

Launch Day plus 6: *Herschel* was now twice as far as the Moon. This was the day chosen to start waking up the instruments.

The first wake-up call was for SPIRE.

Matt Griffin, Bruce Swinyard and several other members of the SPIRE team were waiting at Mission Operations. *Herschel* was now so far away it would take four seconds for a signal to reach the spacecraft and four seconds for a signal to come back.

One of the team pressed a button. Four seconds, a very long breath. Four seconds, another long breath. A signal came back.

Nobody would know whether SPIRE had survived the launch with all its faculties until first light, but at least it was awake.

Launch Day plus 8: The SPIRE team switched on the instrument's helium-3 refrigerator. After one cooling cycle, the detectors reached their operating temperature of 0.29 kelvins.

Launch Day plus 10: The PACS team woke up their instrument. Everything nominal.

Launch Day plus 12: The HIFI† team woke up their instrument. Everything nominal.

Launch Day plus 16: This was a moment that had worried Matt. Moving parts in space instruments are always a worry. One critical part of SPIRE was the moving mirror used in SPIRE's spectrometer, which had been locked into place during launch (Chapter 5). The mirror had already been unlatched. It was now time to test whether it was working by using the spectrometer to observe the inside of the cryostat lid. When someone sent a signal to the spacecraft to make the observation, an oscillating line should appear on a screen at Mission Operations, provided the mirror had moved back and forth correctly. Somebody sent the signal.

Four seconds, a long breath. Four seconds, another long breath.

SPIRE yawned. A wiggle appeared on the screen (Figure 8.2).

Launch Day plus 27: The spacecraft entered its orbit around the second Lagrangian point. If there had been anyone in the spacecraft to see it, the Earth would have appeared about the size that the Moon does in our sky.

Launch Day plus 32: This was the day that *Herschel* would make its first observation. PACS had been chosen as the instrument that would make the first observation. Göran was now at Mission Operations. Albrecht Poglitsch, the PACS PI, and the rest of the PACS team were at the team's headquarters in Munich.

The first step was to open the lid on the cryostat. At the time, I didn't know much of what was going on, but I did know that opening the lid was a worry. The spring that would open the lid had been deliberately designed to be weak so that the instruments would not be shaken too much when the lid opened. If the rumours were correct, opening the lid had only been tested once on earth. The YouTube video of the test was not reassuring. In the video, the lid opened slowly until it was almost completely open, then suddenly flopped back, almost closing, before slowly opening again. If the lid didn't open, we might have working instruments but the only thing we would ever be able to observe would be the back of the cryostat lid.

---

† The Heterodyne Instrument for the Far-Infrared (HIFI) was the third instrument on *Herschel*.

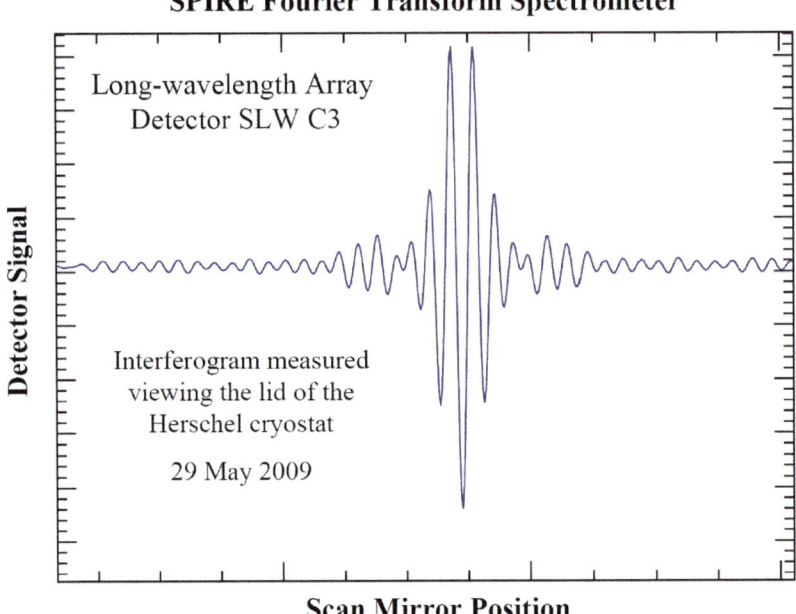

SPIRE Fourier Transform Spectrometer

Long-wavelength Array
Detector SLW C3

Detector Signal

Interferogram measured
viewing the lid of the
Herschel cryostat

29 May 2009

Scan Mirror Position

FIGURE 8.2   The signal received back on Earth after the observation of the inside of the cryostat lid with the SPIRE spectrometer 16 days after launch.

Credit: ESA and SPIRE Consortium

Göran and about 25 people were standing round a screen at Mission Operations when the signal was sent to the telescope. There was no camera on the spacecraft, so they hoped to see a signal come back from the telescope's gyroscopes showing the telescope had moved slightly when the lid opened. Ten seconds – the telescope was now even further away – then the signal came back. The lid was open. The instruments could now see the sky.

The object proposed by the PACS team as *Herschel's* first target was the iconic galaxy Messier 51 (M51). This was the first galaxy shown to have spiral arms[‡] and is sometimes known as the Whirlpool Galaxy. It had been observed in the far-infrared waveband before by the much smaller *Spitzer Space Telescope*, and everyone thought that a more detailed *Herschel* picture shown side-by-side with the old *Spitzer* picture would be an excellent way of showing the promise of the new telescope.

In the original plan, there were to be no observations in the 24 hours after the opening of the lid to allow the temperature of the instruments to stabilise. But then Göran heard from Thomas Passvogel, the *Herschel* Project Manager. The ESA Director General had been on the phone. He wanted a picture now![§]

[‡]   The spiral arms of M51 were first seen with the next Big Telescope after the Herschels': the 72-inch telescope built in Ireland by the Earl of Rosse in the mid-nineteenth century – the 'Leviathan'.

[§]   He wanted a picture to show at the Paris Air Show.

Göran phoned up Albrecht. Albrecht said that even if PACS hadn't quite reached its operational temperature, if they took enough pictures, tweaking the voltages between each one, they might get one decent one. Göran decided to take the train over to Munich so he could be with the PACS team when they saw the first pictures.

An hour out of Munich, while he was working on his laptop, his phone rang. It was Albrecht. He was screaming: 'Göran, we have it! We have it. M51, we have it! It's on the screen. We have it!'. He told Göran, when he got to the station to buy two bottles of champagne, get out to the institute immediately, take a taxi.

When Göran got there, there was a mob of excited scientists and engineers. The old *Spitzer* picture barely showed the galaxy's spiral structure. The picture on the screen revealed the spiral arms in detail (Figure 8.3). To cool the champagne quickly, Albrecht put one of the bottles in a flask of liquid nitrogen. The bottle exploded. Forgetting everything – experimental physics, common sense – Albrecht put the second bottle there as well. It exploded too. Eventually, someone found some warm white wine and they celebrated with that.

I had just spent two days in Paris at another *Herschel* meeting. At the end of the meeting, to enjoy the beautiful spring weather for at least an hour or two, we went for a beer at a pavement café. Someone wandering past said something to an astronomer at one of the tables. The news travelled quickly from table to table. They had opened the lid, everything worked. It was springtime in Paris.⁵

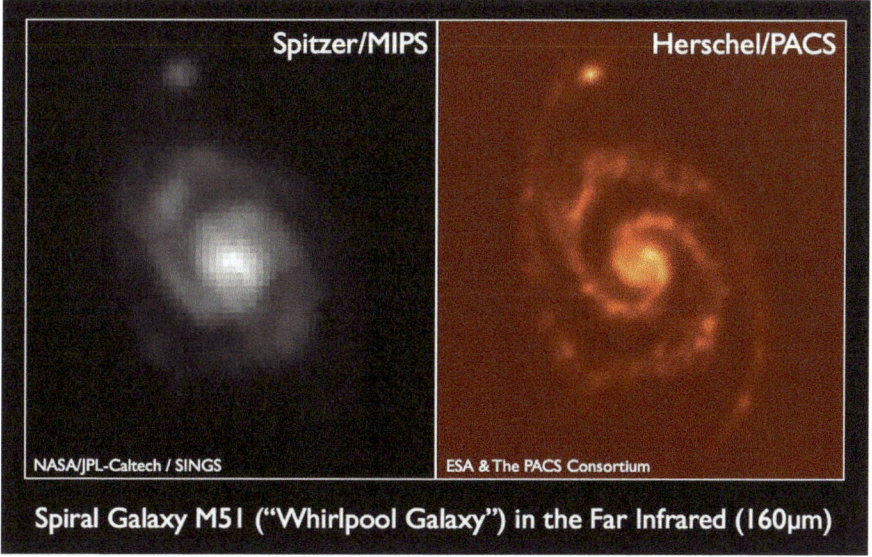

FIGURE 8.3 On the left – the older picture of M51 made with the *Spitzer* telescope; on the right – the *Herschel* first-light picture.

Credit: NASA/JPL-Caltech/SINGS and ESA/The PACS Consortium

---

⁵ I found out later that the person was Laurent Vigroux, co-PI of SPIRE.

Day 42 was decision day for Matt Griffin – first light for SPIRE. The day he would find out the result of ten years of work.

He says he was confident. SPIRE had been woken up successfully. He also started his professional life as an engineer, and engineers expect their designs to work. Of course, they don't usually launch their machines on the top of an Ariane 5. But they are used to designing machines for unique situations, and the ESA engineers had shaken SPIRE almost to destruction on the testbed at ESTEC. Matt trusted the ESA engineers and he also trusted his team. Whatever he says, there must have been some lurking anxieties. He was also a scientist, and scientists look for the unexpected. A scientist can never say they are 100% sure of anything – all swans are white until a black one comes marching over the horizon. And even in the world of engineers, there is the human factor. He trusted his team but suppose one of them had an off-day.

The setting was a nondescript office in the Rutherford-Appleton Laboratory, which the SPIRE team were using for their HQ. The team had chosen Messier 66 (M66) and Messier 74 (M74) as the first-light targets, two other galaxies from Messier's catalogue of things-that-are-not-comets (Chapter 7), not as spectacular as M51 but still beautiful spiral galaxies.

The observations had been made already. A little over a day before, instructions to observe the galaxies had been uploaded to the telescope. Sometime during the last 24 hours, a pattern of electrical currents changed and SPIRE started its first exposure. During the next contact period, the data from the first observations had been downloaded to one of the ESA ground stations, transferred to an ESA centre near Madrid and then over the internet to the Rutherford-Appleton Laboratory.

The data was now on a laptop in front of Matt and some of the key members of the SPIRE team.

One of the most junior members of the team now took charge: Pierre Chanial, a French postdoc at Imperial College. With the cookbook of SPIRE data-reduction programs beside him, Pierre set to work to convert the raw data for M74 into a picture.

Trying not to let the watching scientists unnerve him, he ran the programs, one by one, on the laptop. After about ten minutes, he pressed return to run the final program to make the picture and display it on the screen. The laptop hummed for a few seconds….

No picture. All Matt could see was noise – the snow that appears on a TV screen when there is no signal. Nothing, Rien, Nada….

It was possible that the thing he had not believed would happen had really happened. SPIRE had woken up, but maybe there was something wrong with its software. Maybe something had been jarred loose during launch….

But it was also possible Pierre had made a mistake. For the next ten minutes, Pierre went back through the sequence of programs, patiently checking what he had done. The rest of the group were only able to watch. Pierre tried changing some of the parameters in the programs. No difference. He fiddled with some other parameters….

This time there was something there. On the screen in front of him, Matt saw a picture of M74 (Figure 8.4), the first-ever picture of M74 in the sub-mm waveband.

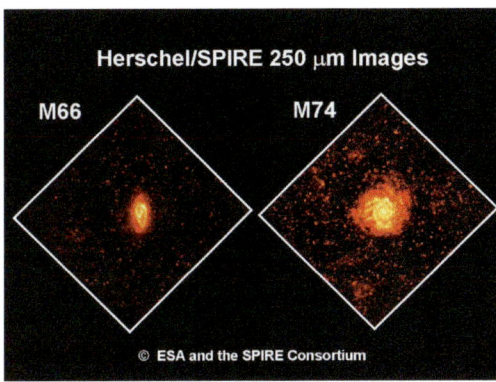 

FIGURE 8.4   Left – the first two pictures taken with SPIRE. The faint red specks around the two bright galaxies are sub-mm galaxies. Right – the members of the SPIRE team who were there at first light. (Matt Griffin is the tall one at the back; Pierre Chanial is the one in the striped shirt sitting by the laptop on which the pictures appeared.)

Credit: ESA/the SPIRE Consortium/Matt Griffin

But he realised immediately there was something else in the picture. The Red Book had promised *Herschel* would detect thousands of sub-mm galaxies, the ancestors of present-day elliptical galaxies. These objects were the reason Matt and Bruce had designed SPIRE the way they did (Chapter 5).

And Matt could see them now! Around M74, a galaxy that in cosmic terms we are seeing only yesterday,** he could see faint blobs, which he realised must be sub-mm galaxies, sub-mm radiation from galaxies billions of years in the past.

This picture, the first ever picture with SPIRE, was also a core through the history of the universe.

The success of the first-light observations made the next three months frustrating for astronomers like me. This was the period of Performance Verification, the time set aside for testing all the observing modes of the three instruments. We knew they were working, but we were not able to use them yet. We had to try to wait patiently while the engineers and the instrument teams checked everything out.

It was less frustrating for the astronomers who were key members of the instrument teams. Mike Barlow was a member of the SPIRE team and a professor at University College London, close enough to the Rutherford-Appleton Laboratory that he could easily get there and watch the observations come in. Matt Griffin had asked Mike to suggest some targets to test the spectrometer part of SPIRE. Mike suggested supergiant stars and planetary nebulae.

Planetary nebulae, which are possibly the most beautiful objects in the universe, may have been another of the things discovered by William Herschel. He does seem to have thought of the name, although it is a misleading one because they are nothing to do with planets.[1] Compared to most of the objects in the universe, they have very ephemeral lives,

---

**   About 30 million years ago, which, since the universe is 14 billion years old, is only yesterday in cosmic terms.

only about 10,000 years. And they have something of the transience and beauty of flowers, although that comparison is not quite right because every planetary nebula looks very different from every other one. To me, they look a little like the strange translucent creatures discovered in the deep ocean (Figure 8.5).

A supergiant and a planetary nebula are the pre-death and post-death phases of a star. A star begins to die when its core runs out of hydrogen. As the nuclear fusion reactor in its core shuts down, the outer layers of the star swell and the star becomes a red giant. Eventually, the temperature in the core becomes high enough that nuclear fusion restarts – this time with helium rather than hydrogen as the fuel (Chapter 13). But eventually, the core runs out of fuel again. The star now swells up even more, becoming a supergiant. If the star has

FIGURE 8.5 The most beautiful objects in the universe – planetary nebulae. Clockwise from top-left: the Butterfly Nebula, the Catseye Nebula, the Spirograph Nebula and the Little Gem Nebula.

Credit: NASA and the Hubble Heritage Team (STScI/AURA)

a low mass, it is now in its terminal phase. It is now seriously unstable, its swollen atmosphere is no longer held securely by gravity and there is a steady flow of gas away from the star. Eventually, the star gets so unstable that all its outer layers are ejected, exposing the hot core. The intense radiation from the exposed core (now with a pretty new name: a white dwarf star) irradiates the ejected gas. The irradiated gas is the planetary nebula. As the gas gradually disperses, the flower fades.

During the supergiant phase, the star becomes a vast chemical factory. Elements made by the fusion reactions in its core circulate through its swollen atmosphere. The gas in a star is usually too hot for molecules to exist, but the outer layers of the supergiant are cool enough for molecules like carbon monoxide and water to be formed. It is even cool enough for solid particles – dust grains – to form a smog within the gas. After the gas has been ejected, its irradiation by the white dwarf makes a different set of chemicals.

Mike knew that supergiants and planetary nebulae must be packed full of interesting chemicals. He suggested they would be the perfect targets to test SPIRE's spectroscopy mode.

After the spectra began to appear on the screens at the Rutherford-Appleton Laboratory, Mike and the others couldn't believe how many spectral lines they could see. In the spectrum of the supergiant star VY CMA (Figure 8.6) there are lines everywhere, lines upon

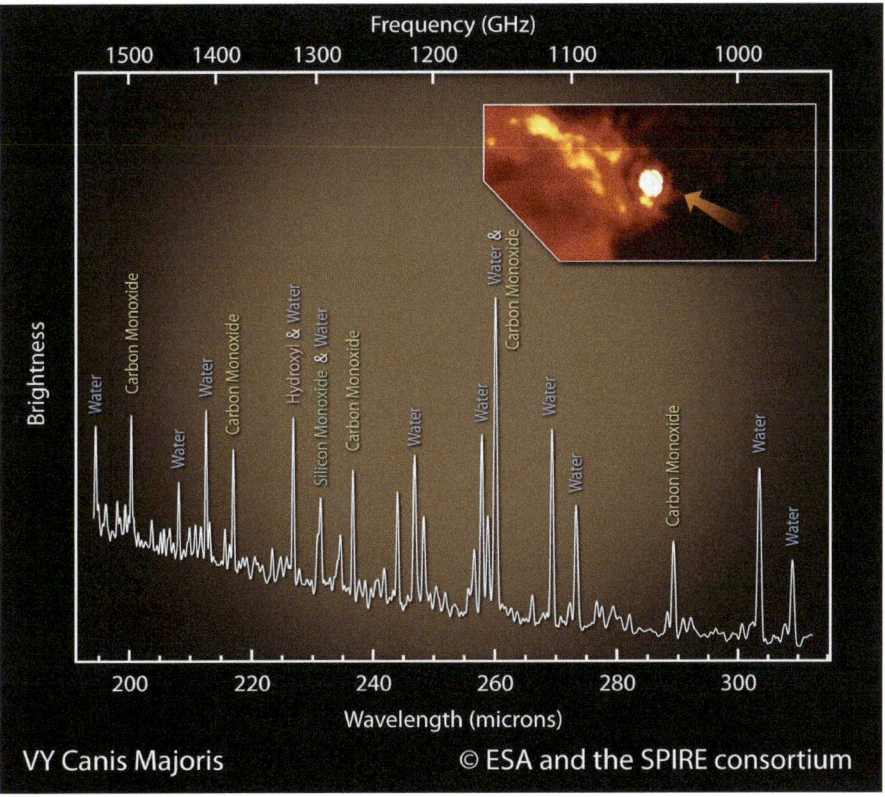

FIGURE 8.6   A small part of a SPIRE spectrum of a supergiant star.

Credit: ESA/the SPIRE Consortium

lines, from spectral lines emitted by common molecules, like carbon monoxide and water, to spectral lines from chemicals such as formyl, cyanogen, methylidine and ethynyl, which are almost never seen on earth because their molecules are so reactive. And in those first spectra that appeared on the screens, there were also spectral lines nobody could even identify.

When faced with an unknown spectral line, an optical astronomer looks in a book; there are reference books that contain laboratory measurements of the wavelengths of virtually all the spectral lines that fall in the optical waveband. But this was a new waveband and many of the lines in the first SPIRE spectra had never been seen before, either in space or in a laboratory. The person with the most experience of the sub-mm spectra of stars was another member of the SPIRE team, the Spanish astronomer Pepe Cernicharo. Pepe has spent his career looking at the spectra of interstellar chemicals, but when he saw the first spectra from SPIRE, even he went 'Wow!'. A year later, even with Pepe's help, there were many lines in the SPIRE spectra that nobody had been able to identify.[2]

For some astronomers, Performance Verification was therefore not the boring time while we waited to use the telescope. Years later, many of these first spectra are still being used to investigate the exotic chemistry of interstellar space.

During Performance Verification, the *Herschel* command centre was a special visitors room at the Mission Operation Centre. Every morning at 11 o'clock, Göran was there, ready to talk to the people in the instrument teams. The mood depended on what had happened during the last contact period.

On a good day, when the first messages came from the spacecraft, the status of all the instruments would be nominal. The teams would then spend the first hour and a half downloading the data from the last observing period. They would then upload the instructions for the next two observing periods, during which *Herschel* would sweep across the sky, moving from object to object – a robot astronomer observing the universe.

On a bad day, one of the instruments would be down. This usually happened because of a single event upset (SEU), in which some position in the instrument's computer memory had been hit by a cosmic ray. The agreed rule was that the instrument then had to stay down until the instrument team, doing some rapid diagnostic tests during the contact period, had figured out which bit of the memory had been corrupted and showed it was safe to turn the instrument on again. If they didn't manage to do this, the badness of the day would depend on the telescope's schedule. If SPIRE was down and it was next on the schedule (with uploaded instructions ready to go), the next observing period would be lost. But if it was SPIRE that was down and PACS was next on the schedule, the observations could still go ahead – and if the SPIRE team had figured out what was wrong before the next contact period, no observing time would be lost. There were many more good days than bad days, but overall three per cent of the total observing time was lost.

The 81st day after launch, Operational Day 81,[††] was a very bad day.

---

†† Operational Days started at launch, but no observations of the sky were made until the opening of the cryostat on Operational Day 32.

Unlike the other two instruments on *Herschel*, which were able to take both pictures and spectra, HIFI was a specialised instrument designed to carry out forensic spectroscopic investigations of interstellar gas. It had been woken up successfully on Operational Day 39 and had made its first observations, but when the contact period arrived on Operational Day 81 the *Herschel* team found it was down. They realised quickly this was not just an SEU, some bit of memory knocked out by a cosmic ray, but something much worse. There had been a voltage surge in the instrument which had burnt out one of the diodes[‡‡] in the electronics. HIFI was broken.

This was not necessarily a disaster. HIFI did have a set of backup electronics. But there was no point in turning on the backup electronics in case the same thing simply happened again. The HIFI team at Groningen in the Netherlands and the Polish team that had built the electronics started to work like crazy, supported by work in the labs at ESTEC, trying to figure out what had happened. The initial tests did not look good. It looked as if there was a problem with the original design. Some of the diodes seemed to be operating at voltages outside their safe ranges. If this was true, HIFI might operate for a couple of months when the reserve electronics was turned on, but then the same thing would probably happen again. Anyway, for now, HIFI was out of action.

At the end of the summer, Performance Verification ended. The Science Demonstration Phase began.

[‡‡] A diode is a component in an electrical circuit that only lets the electrical current flow in one direction.

CHAPTER 9

Into the Rift

THE SCIENCE DEMONSTRATION PHASE, the SDP, was one of Göran's ideas. It would last from October to December, during which each of the key-project teams would get some early observations, which we could use to test our observing plan and get some early results. In return, we would need to present our results at two conferences and write some quick science papers about them. The two conferences would be in Madrid in December and at ESTEC the following May, with the deadline for the papers just before the ESTEC conference. Göran hoped that the conferences and the papers would generate a big bang of publicity for the telescope.

My only problem with the idea was that the schedule for the papers did seem very challenging. I suspected that Göran hadn't written a paper himself for a long time and had forgotten how long it can take. It can often take years to analyse the data for a project and write the papers describing the results, but Göran's schedule meant that a team whose observations were made in December would have only five months to analyse the data and write the papers. On the other hand, the prospect of getting some early observations was very exciting.

By November, my excitement had waned quite a lot. The team that Loretta and I had assembled for our big survey, the *Herschel* ATLAS, had still not received any data. It was now six months since the launch of the telescope. Who knew whether the thing that had happened to HIFI might also happen to PACS or SPIRE! While we waited, *Herschel* made its first big discovery.

It was William Herschel – of course – who was the first person to try to understand how a star is born. He thought he had disproved his own grand theory that the nebulae are all individual galaxies (which we now know to be mostly true) when he noticed that there were some nebulae that seemed to be associated with individual stars (Chapter 7). But this disappointment inspired another of his daring conjectures – that the star might have been created by the contraction of the nebular fluid as the result of gravity.[1] Some of his

DOI: 10.1201/9781003195290-9

conjectures were a little off – the people on the Sun, for example – but this was one of the better ones.

Two centuries later, though, it sometimes seemed that we hadn't got much further than him in understanding the birth of a star. We do now know a few more things. We know that the stars are born in the huge clouds of molecular gas – the giant molecular clouds. We also know that he was basically correct, that it is the contraction of gas within a cloud, as the result of gravity, that leads to the birth of a star, and we also understand this process a little better.

Somewhere deep inside a molecular cloud, the gas becomes dense enough to collapse under its own weight. As the gas accelerates inwards, the energy it initially had from its position in the cloud's gravitational field – like the cup balanced on the edge of my chair in this coffee shop – is converted first into the energy of motion, kinetic energy, and then, as the collapsing gas jostles together, into heat (or, in the case of the cup, if it falls on the floor, into a shower of fragments). When the gas gets dense and hot enough, nuclear fusion starts, energy is released, which increases the pressure of the gas, which stops the collapse – and a star is born.

But that's not that much more detail than in William Herschel's original proposal and there are still missing parts of the story. The most embarrassing gap is that we don't understand why all the gas doesn't immediately collapse.

Although the gas in molecular clouds is very cold, usually about 10–20 kelvins, it does exert a pressure, the same kind of pressure that is in the air in a bicycle tyre. This pressure, like the pressure in the tyre, resists the force of gravity. The big problem, though, is that this pressure is nowhere near enough to stop a giant molecular cloud from collapsing. If the pressure in the gas was the only thing resisting gravity, every giant molecular cloud in the Galaxy would immediately collapse and form stars. The clouds are not collapsing, so there must be something else resisting gravity, but although there are a couple of possible explanations, there is not yet any clear evidence which is correct.

There is one reason above all why in 200 years our knowledge of how a star is born has advanced so little: interstellar dust. Gas and dust are always found together, and the dust shrouds everything in a giant molecular cloud, hiding the births of stars almost completely from optical telescopes, even one like *Hubble* that seems to have taken spectacular pictures of just about everything else in the universe. And this, of course, was one of the reasons why everyone was so excited about the launch of *Herschel*. Its importance was that, as a sub-mm telescope, it gave us a way of seeing through all this dust.

Suppose there are some newborn stars deep inside a giant molecular cloud. Even better, let's suppose they are protostars. These are the dream objects for people who work in this field because these are dense hot regions of gas that are still contracting but are not yet hot and dense enough for nuclear fusion to start. They are on the way to becoming stars but are not quite there yet. Observing them would tell us a lot about the pre-birth development of a star. Of course, the protostars are hidden by the dust, and there is no way with an optical telescope we would even be able to tell they are there.

The dust close to the protostars, however, is a littler warmer than the dust in the rest of the cloud, and thus a stronger source of far-infrared and sub-mm radiation. Unlike visible light, this radiation passes straight through the dust in the rest of the cloud. And so observations with a sub-mm telescope are a way of finding all the newborn stars and protostars deep inside a molecular cloud, something that would never be possible with an optical telescope like *Hubble*.

This had been one of the main themes of the Red Book, and there were many astronomers queuing up to use *Herschel* to look for the protostars and newborn stars inside molecular clouds. Within Matt Griffin's SPIRE team, the group given the job of planning the observing programme to investigate the birth of stars was Specialist Astronomy Group 3, or SAG3 as everyone called it (Chapter 6). One of the leaders of SAG3 was Philippe André, who works in the group of astronomers at the Atomic Energy Commission in Paris. Even in a community full of intense individuals, Philippe is known for his intensity. For the last few decades, he has focused this intensity on trying to understand how a star is born.

Philippe and the group decided to use all their observing time to take pictures of the closest molecular clouds. All the closest clouds happen to lie in a ring centred roughly on the Sun, about 3,000 light-years across and at an angle of about 15-to-20 degrees to the Milky Way. The Gould Belt, named after the American astronomer who discovered it in the nineteenth century, is one of astronomy's many casual mysteries - nobody understands why it exists, why it is centred on the Sun, and why it is at an angle to the Milky Way. Philippe and the group decided to take pictures of all 15 molecular clouds in the Gould Belt with both *Herschel* cameras, PACS and SPIRE, which would give them pictures of each cloud at five wavelengths.

Like all the other teams, they had been given a small amount of time in the SDP for some early observations. There was just enough time to observe two clouds, so they had to choose which two. They picked ones at opposite extremes.

Their first choice is not actually a molecular cloud. The molecules in molecular clouds are mostly molecules of hydrogen, two atoms of hydrogen spliced together. The Polaris Flare, though, a plume of gas that rises out of the Galaxy's disk, is mostly made of hydrogen atoms. Philippe and the group chose it as an example of a cloud of gas where they didn't expect to find many newborn stars or protostars.

At the other extreme was the Aquila Rift, which gets its name because, in the visible waveband, it appears as a winding dark canyon outlined against the light of the Milky Way. The darkness is caused by dust, and since gas and dust always go together, the team hoped they would find many protostars and newborn stars there. They may also have been drawn to it by its sense of mystery. The Rift is tantalizing and it's not surprising that it's been the subject of at least one science-fiction book,* although, as a Star Trek fan, I am disappointed the USS Enterprise has never been there.

On Operational Day 162, a pattern of currents shifted and *Herschel* began observations of the Polaris Flare. A day later the pattern of currents shifted again. The telescope began observations of the Aquila Rift.

When Nico Peretto, a postdoc in Philippe's group, saw the first picture of the Polaris Flare appear on the screen (Figure 9.1) his first thought was 'there are filaments everywhere'. In later *Herschel* pictures, the filaments are even more obvious (Figure 9.2). *Herschel's* first

---

* *Beyond the Aquila Rift* by the British sci-fi writer Alastair Reynolds, now a *Netflix* series.

FIGURE 9.1   The *Herschel* picture of the Polaris Flare.

Credit: ESA/Herschel/SPIRE/Gould Belt Key Project

big discovery, which jumped out of the screens during the SDP, is that the Galaxy is full of filaments, tendrils of co-mingled gas and dust that snake through interstellar space.

An obvious question is: why didn't we know about these filaments before the launch of *Herschel*?

It is true that we have been using observations of the CO spectral lines to map the molecular gas in the Galaxy for almost 70 years now (Chapter 5). But the answer to the question is that the CO method has some serious flaws. Almost all the molecular gas in the Galaxy is molecular hydrogen, which is impossible to detect directly. There is roughly one CO molecule for every 10,000 hydrogen molecules, which is why observations of the spectral lines from CO, which are easy to observe, have been used to trace the invisible molecular hydrogen. But the first problem with the CO method is that there is evidence that in some places in the Galaxy the molecular hydrogen does not contain any CO at all.[2] The second problem is that the strength of the CO spectral lines are an unreliable guide to how much molecular hydrogen is present.[†] Filaments had been seen before, but these flaws with the CO method had concealed their importance.

Until we saw these pictures, astronomers described the interstellar gas and dust as being in clouds, but these pictures seem to suggest that it is better to think of them as being in filaments. But the truth is that we are limited by our vocabulary. When we try to describe

[†]   The most widely used CO spectral line is so strong that above some critical amount of CO the spectral line saturates, with any further increase in the amount of CO not producing an increase in the strength of the spectral line (the extra radiation is absorbed by the CO itself). One of the many advantages of dust over CO[2], as a way of tracing the invisible molecular hydrogen, is that twice as much sub-mm radiation unambiguously means there is twice as much dust.

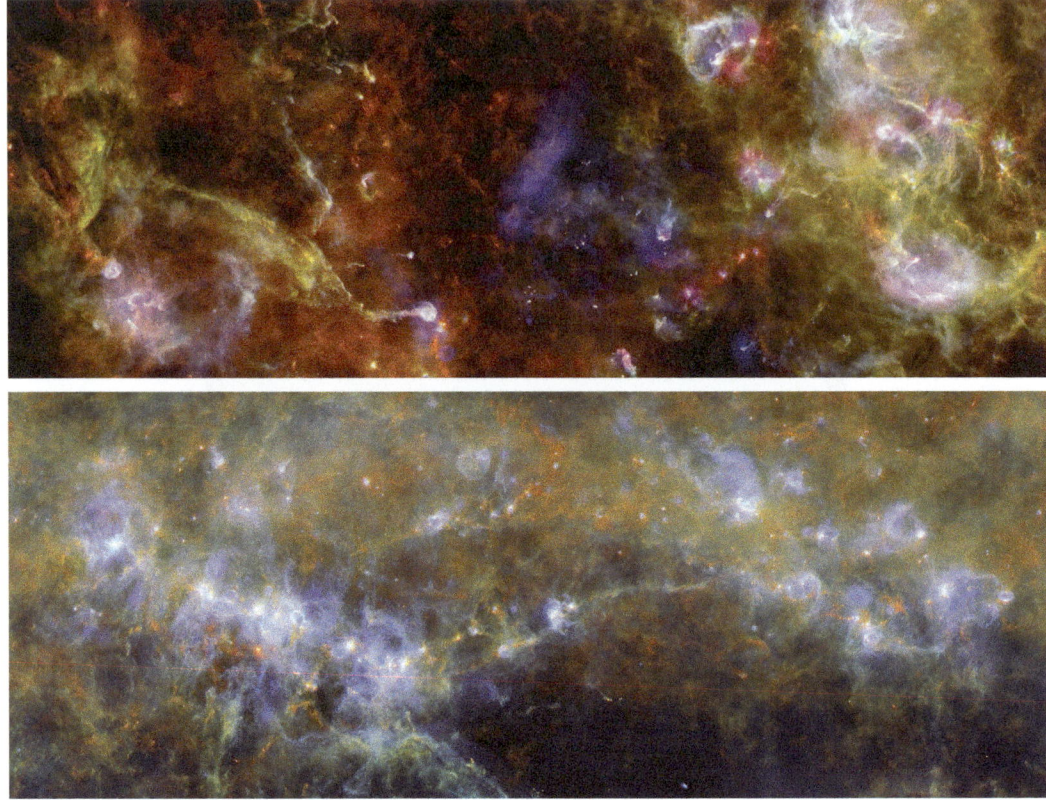

FIGURE 9.2   More filaments!

Credit: ESA/Herschel/PACS,SPIRE/Hi-GAL Project, HOBYS Consortium

the forms we see in the interstellar gas and dust, we necessarily reach for words from our own natural world. The structures in the CO maps do look a little like clouds and the structures in the *Herschel* pictures look more like filaments or tendrils, but the structures we see in the interstellar gas and dust are *sui generis*, in a class by themselves. As it happens, I think the picture of the Polaris Flare looks like the pattern produced when ink is injected into water, and some of the things in the other pictures look to me like the neurons seen in brain scans. But let's stick to clouds and filaments for now.

This first discovery jumped out of the screens. The second discovery made by the group required some careful measurements.

The advantage of taking pictures at five wavelengths is that it is possible to combine the images to estimate the temperature of the dust – the higher the temperature, the stronger the radiation detected in the short-wavelength images (Chapter 2). Philippe and the group combined their images of each molecular cloud to make a single picture, using colour to show the temperature of the dust – white where the dust is hot, blue where it is a little colder, red where it is colder still (All the pictures I have shown in this chapter have been made this way).

FIGURE 9.3   A study in colour – the first *Herschel* picture of the Aquila Rift. The colour shows the temperature of the dust, the temperature descending through the colours: white, blue, yellow and red.

Credit: ESA/Herschel/SPIRE/PACS/Gould Belt Key Project

When they looked at their picture of the Polaris Flare (Figure 9.1), they saw a study in beige. The colour, and so the temperature of the dust, is the same everywhere, which showed, as they had expected, that there are no protostars and newborn stars to produce warm spots in the dust.

But when they saw the first picture of the Aquila Rift (Figure 9.3), it looked as if it had been painted by William Turner on one of his more exuberant mornings. Although the filaments are less obvious here, camouflaged by the spectacular colours, the team found, when they inspected the picture carefully, that it is still threaded by a network of filaments.[3] The dabs of white along the filaments, which can be seen most clearly in the top left of the picture, are warm places in the dust, the positions of newborn stars and protostars. This picture and the other pictures (Figure 9.2) show that it is the filaments in which the stars are born.

But there are also filaments in the Polaris Flare. Philippe and the group were faced with the question of why there are protostars in the Aquila Rift but none in the Polaris Flare. They found the answer with some careful measurements.

Using the strength of the sub-mm radiation from each filament to estimate how much gas and dust it contains[‡], they discovered that in the Aquila Rift it is only in the densest

---

[‡]   They were able to estimate the mass of the dust in a filament from the strength of the sub-mm radiation, using its relative strength on the images at the five different wavelengths to correct for the effect of the temperature of the dust on the strength of the radiation (warm dust radiates more strongly than cold dust – Chapter 2). On the reasonable assumption that the ratio of the mass of gas to the mass of dust is always roughly the same, they were able to estimate the combined mass of the gas and dust in the filament.

filaments where there are any protostars. There seemed to be a critical density: if the mass of each light-year of filament was greater than five times the mass of the Sun, there were protostars in the filament; if the mass per light-year was less than this, there were no protostars in the filament.

When they looked at the filaments in the Polaris Flare, they realised this was also the explanation of why there were no newborn stars or protostars there. None of the filaments there had a density greater than the critical density.

And then Philippe remembered a paper he had read by two Japanese astronomers in the 1990s. The two astronomers, theorists, had used various bits of physics to investigate whether a filament of gas in interstellar space would be stable. They had predicted that if the density of the filament was above a critical density, it would collapse as the result of the internal gravitational forces. When Philippe looked up the paper, he discovered that the predicted critical density was almost exactly what his team had found.[4]

From these early observations, Philippe and the group had filled in one of the missing pieces of the story. The birth of a star, at least a low-mass star like the Sun,[§] occurs in a filament of co-mingled gas and dust. The birth starts when the density of the filament exceeds the critical density. The filament becomes gravitationally unstable and starts to collapse. The filament breaks up into a string of dense heaps of collapsing gas. The gas heats up; the density of the gas increases; nuclear fusion starts....

And then they discovered something else interesting. Not with any definite scientific motive but because it was something easy they could do, they had measured the widths of all the filaments. They discovered something surprising – that all the filaments, whether in the Aquila Rift or the Polaris Flare, had almost exactly the same width: 0.3 light-years. Eventually, someone – nobody remembers who – suggested it might be something to do with turbulence.

I didn't mention above the possible explanations of why all the giant molecular clouds don't immediately collapse to form stars, but one of them is turbulence.[¶]

Turbulence is found everywhere – in the atmosphere, the oceans, even in my cup of coffee – so it would not be surprising if it is also important in interstellar gas. Turbulence starts when something stirs up a liquid or a gas. If I stir my coffee, the coffee starts by swirling around my cup. The swirl breaks into smaller swirls, which then break into even smaller ones – in no particular pattern, this is a chaotic process – and eventually, after a few seconds, the swirls are so small they are no longer visible. The churn-up in the interstellar gas and the vigorous motions created by turbulence are one possible explanation of why all the giant molecular clouds in the Galaxy do not immediately collapse under their own weight.

The SAG3 team began to suspect that turbulence might also be the cause of the filaments. The *Herschel* pictures looked to them remarkably like places in our own natural world where turbulence is playing an important part – in the structures of the clouds in our atmosphere, for example. In interstellar gas, though, turbulence behaves in a different way

---

§   The stars being born in the Gould Belt are mostly low-mass stars like the Sun.
¶   The other main suspect is the pressure in the magnetic fields that thread through the interstellar gas.

from the way it behaves in the Earth's atmosphere. In such a low-density gas, some of the gas buffeted by the turbulence moves at supersonic speeds,** which should generate shock waves. These shock waves should sweep up the gas and dust, leaving dense heaps in their wake. It seemed to the team that the filaments might be these heaps. Then somebody remembered that in one version of the turbulence theory, the width of a heap should always be the same, no matter the speed of the shock. When they went back to look at the relevant papers, they realised that the predicted width of a heap was just about the same as the width they had measured for the filaments: 0.3 light years.[5]

The filaments in the *Herschel* pictures are therefore probably produced by interstellar turbulence. It seems likely that the filaments in the Polaris Flare are transient structures – interstellar weather. It is only when the density of the filament exceeds the critical density that gravity takes over, the filament collapses and stars are born.

One question they had not answered, though, is what generates the turbulence. What is the spoon in the coffee?

To me, the answer leaps out from the *Herschel* pictures. There are loops, whirls and arcs everywhere – each arc showing a place where there is probably an unseen star (the stars are invisible because, compared to the dust, they emit hardly any sub-mm radiation) injecting energy into the gas and dust around it. Whether it is the energy in the jets and radiation from stars at the beginning of their lives or the winds and explosions at the end of their lives, I don't know – and I can't prove any of this – but it seems likely to me that it is the stars themselves that are generating the turbulence. It is the stars that are making the weather. William Turner tried so hard to capture the turbulence in the Earth's atmosphere that it seems appropriate many of the *Herschel* pictures look like his paintings.

But when I look at pictures of the Aquila Rift, there remains for me that nagging question: What lies beyond? The USS Enterprise never went there, but if it did….

"Sulu, follow that ship!"

As we descend into the Rift following the Klingon ship, we start to see wisps of dust. The dust thickens. We see veils, loops, tendrils, rivers, columns and mountains of dust. The stars gradually dim, although sometimes, suddenly, we pass through a gap between two cliffs of dust and see the stars again as bright as ever. We never lose the stars completely, though, and the dust is not thick enough for us to lose sight of the Klingon ship. For days, we follow it into the Rift. Those of us on the bridge can see it on the big screen. It must be aware that we are following it, but it makes no attempt to lose us.

Then one day, ahead of us we see a huge serpent of gas and dust writhing through the Rift. After a few hours, its dark surface fills the screen. The Klingon ship makes a sudden dash for it. The captain orders the Enterprise to follow. Once we are inside, the stars vanish completely. The screen on the bridge is empty. The Klingon ship can't escape, though,

---

** Sound waves, which are just pressure waves, travel through interstellar gas as they do through our atmosphere. Supersonic speeds are speeds that are faster than the speed of the sound waves.

because our ship has infrared sensors, which allow us to follow it through the dust by detecting the heat of its engines. We follow it for many days as it zigs and zags to stay within the serpent. With the loss of the stars, everyone's mood has darkened. Interstellar spaceflight is never easy and the stars were our last reminder of home.

Then one day the sensors detect a second infrared signal. The signal slowly increases in strength until it is much larger than the signal from the ship.

And then everything suddenly goes crazy. The screen is filled with light. The ship lurches from side to side. The sensors show the temperature is climbing rapidly. Someone panics, shouting we are inside a star.

Our Science Officer – it's probably his half-Vulcan ancestry – doesn't panic. He realises the ship's sensors are not detecting any neutrinos, which means we can't be inside a star. We must be inside a protostar. This is still potentially lethal – life support on the Enterprise will fail if the temperature goes too high – but we have a little more time to figure things out. While he gets on with his job in his methodical Vulcan way, the rest of the crew goes crazy. The temperature keeps on rising, people pass out on the bridge and fights start.

Eventually, Spock works it out. The radiation must be strongest in the direction of the centre of the protostar. We set course in the opposite direction. We burst out of the protostar and, after a short turbulent voyage, emerge from the Rift – on the other side.

We are now the first ship to have gone beyond the Rift, but we are still only one-tenth of the way to the centre of the Galaxy. Shortly ahead of us is the start of the vast Scutum-Centaurus spiral arm. In the distance, we can see the Galaxy's central stellar bulge and, looking away from the disk, we can now see a few galaxies, the closest in a universe of two trillion galaxies.[6]

This is what our survey, the *Herschel* ATLAS, was meant to study. We were still waiting for our observations.

CHAPTER 10

Data Monkeys and Cooler Burps

B Y LATE NOVEMBER, THE news from HIFI was a little better. There was a set of backup electronics in the instrument, but the HIFI team couldn't simply turn it on because they were worried the same thing might happen again. But there was also a spare set of instrument electronics on the ground. When the scientists in the HIFI team in Groningen and the Polish electronics team did some tests, they discovered that although the diode had been operating outside its safe voltage range, it now looked as if that range had been too conservative; the voltages had not been high enough to explain why the diode had burned out.

The real problem seemed to have been a switch. The switch had been designed to protect the electronics from a voltage surge. It was supposed to be off when HIFI was rebooted and only switched on when observations were about to start. On the reboot on Operational Day 81, however, the switch had been left on, which had led to the voltage surge that had burnt out the diode, which led to the months of delay and a traumatised instrument team. The obvious solution was to reprogram HIFI so that the switch was always set to off during a reboot. This should work – if the engineers were right that the diode's normal operating voltages were safe.

At least there was a plan.

On 22 November 2009, Operational Day 192, *Herschel* finally observed one of our fields. We had been allocated 600 hours for our survey, the *Herschel* ATLAS, and Göran had given us 16 hours for our taster session during the Science Demonstration Phase. We needed to think very carefully about how to use them.

We were planning to observe five fields in our survey: two big fields near the north and south Galactic poles and three smaller ones on the celestial equator.[*] For the Science Demonstration Phase the big fields were out. A space telescope is a ponderous thing. *Herschel* could rotate fully around only one axis, and it was also important to avoid pointing it too close to the Earth or the Sun, which meant it was only possible to observe any object during

---

[*] These are described in Chapter 6.

DOI: 10.1201/9781003195290-10

limited periods – 'visibility windows'.[†] The big fields were in awkward places in the sky with very few visibility windows, and since the Science Demonstration Phase would last only a few months we decided to leave these fields for later. The visibility windows for the three small fields were longer, and we also had another reason for preferring them.

A sub-mm survey is only the first step. A list of sub-mm sources, places in the sky that are sources of sub-mm radiation, by itself doesn't tell you much. The vital next step, without which any useful scientific results would be impossible, is to identify the objects – stars, galaxies, quasars – that are emitting the radiation. And for this, we needed an optical image, a picture. And if the object turned out to be a galaxy or quasar, we needed its redshift, which is the essential thing for calculating everything important about a galaxy: its distance, how far back in time you are looking, its luminosity, its mass, etc. The other reason for preferring the small fields was that there were already optical images of these, and there were also redshifts for tens of thousands of galaxies in these fields because of the spectroscopic survey being carried out by another team (Chapter 6). We thought all this extra information would give us a head start on the science, and since it was now late November, only five months until Göran's crazy deadline for the first papers, we thought we might need it. With only 16 hours, our sub-mm image would cover only one-quarter of one of these fields, but it would still be the largest sub-mm image of the sky ever made.

On Operational Day 192, *Herschel* started our first observation. During the next contact period, the data was downloaded to one of the ESA ground stations, transferred to the European Space Astronomy Centre outside Madrid and then by one of our team over the internet to Cardiff.

Many books about astronomy start with the finished products of a survey: the images and the catalogues. They describe the discoveries made with these but skip over the messy business of how these images and catalogues were made. But astronomers generally spend little time on this final stage, the stage of scientific discovery. Most of our time – and it was certainly true in our team – is spent in making the data products. Most of this chapter is about how our team did this. There is an important scientific discovery at the end of the chapter, but this is the story of how we got there. This pre-science stage often takes years, but because of Göran's deadline, we had to complete this stage for our first batch of data in less than a month.

I didn't do much of it myself. Up to the time of *Herschel*, I had always liked to reduce some of the data myself, but with the mammoth datasets that were now arriving for our survey this just wasn't possible. The real work on the data over the next eight years was

---

[†] To avoid consuming too much energy, the direction in which a space telescope is pointing is changed by using a massive spinning 'reaction wheel'. By transferring energy back and forth between the wheel and the telescope, the telescope can be rotated around the wheel's axis. The downside of this method is that since the telescope can be rotated around only one axis, it is only possible to observe part of the sky at any time, but since *Herschel* was also travelling around the Sun, every object would eventually come into view. Any object or field could therefore only be observed by *Herschel* during limited visibility windows.

done by a team of young astronomers. In the beginning, the team in Cardiff was my post-docs, Robbie Auld and Simon Dye, another postdoc who worked for Matt Griffin, Michael Pohlen, my PhD student Matt Smith, and a new young lecturer in Cardiff, Enzo Pascale. Over the next eight years, the group changed – people left to take other jobs and we hired new people to replace them – but the conversation stayed the same, sprinkled with terms nobody outside the *Herschel* data-reduction groups would understand: glitches, jumps, turnarounds, drifts and, everyone's favourite, cooler burps.

The first data monkey was probably Caroline Herschel herself, who after a night's obser-vations would spend much of the day calculating the positions of the nebulae she and her brother had discovered during the night (Chapter 7). Today, we use computers and run a sequence of programs to transform the raw data into a finished image or spectrum, but in essence it is just the same. I have always enjoyed it. There is a lot of pleasure in doing something methodical that requires just the right amount of thought and creativity, and there is also the excitement of the big reveal in which you suddenly see the finished image or spectrum on the screen.

But now I was not doing any of this. It was all the responsibility of the data-reduction team. They were not starting from nothing. Before launch, programmers in the instrument teams had spent years writing the data-reduction programs. But when it arrives, the real data never looks quite like the perfect data envisaged when the programmers were writ-ing the programs. The job the team now had was to adapt these programs, and sometimes write new ones, to deal with the real data with all its messy idiosyncrasies. We had not wasted the six months while we had been waiting for our observations. During that time, our team had been testing their methods and programs on some of the data that had been taken over the summer to check out SPIRE during Performance Verification (Chapter 8).

‡A sub-mm picture may look like an optical picture, but it is made in a completely differ-ent way. The detectors in SPIRE were bolometers, essentially sensitive tiny thermometers§ which register a temperature change when radiation falls on them – very different from the solid-state detectors in an optical camera. We also take the picture in a different way. Our pictures are taken by scanning the camera across the sky, with each bolometer register-ing a temperature change when it passes across a galaxy or some other source of sub-mm radiation. One of the most important data-reduction programs is the one that takes the temperature records from the bolometers – the 'timelines' we call them – and transforms them into a sub-mm picture.

When the team looked at the timelines from the SPIRE data taken over the summer (Figure 10.1), there were lots of surprises. The ideal bolometer timeline would be a fuzzy horizontal line, the fuzz being the noise that is the inevitable part of any astronomical observation, with the occasional peak where the bolometer has passed over a source. But the team often saw places where there was a sudden jump in the timeline, as there is in

---

‡   If you don't find the techniques of sub-mm astronomy fascinating, I suggest skipping the next page and a half and
    starting at the paragraph that begins 'The picture…'.
§   They are very, very sensitive. If the energy absorbed by a bolometer from the sub-mm radiation from the Orion
    Molecular Cloud, which is one of the brightest sub-mm sources and very easy to detect, was used to boil an egg it
    would take about 300 million years.

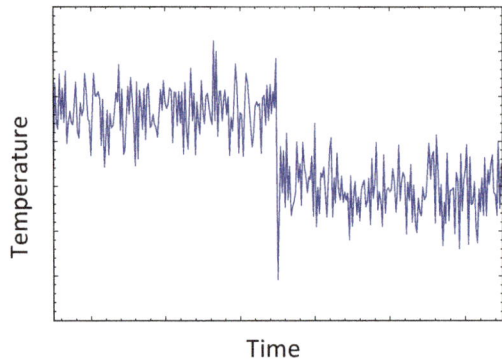

FIGURE 10.1   Part of the temperature record of one of the SPIRE bolometers – the 'timeline'. This particular part does not contain any signal from a galaxy. The fluctuations are the noise and the 'jump' in the middle is one of the things the data-reduction team had to figure out how to remove from the timeline.[1]

Data Monkeys and Cooler Burps

the figure. We never understood the cause of these jumps, but the team had to find a way of removing them from the timelines. The team also saw places where the signal suddenly went crazy, 'glitches', which everyone assumed were caused by a cosmic ray hitting the bolometer. The real timelines were also rarely flat. This was not unexpected because although bolometers are very sensitive, they are not very stable; a real timeline tends to drift up and down because of changes in the bolometer's own temperature, changes that have nothing to do with any astronomical signal. To account for this, SPIRE also contained tiny thermistors, devices that kept a record of the actual temperature of the bolometers, but the team still had to figure out a way of using the thermistor measurements to correct the timelines. The team, led by Enzo, worked out ways of correcting all these problems. There were also times when the signal changed in a very peculiar way, which we eventually realised always occurred six hours after a cycle of the SPIRE refrigerating system. Robbie Auld named these 'cooler burps'. The team invented a way of fixing these as well.[2]

By the time we received our own observations, the data-reduction team was ready to go. Simon Dye and Robbie Auld, who reduced the data, took only a few days to fix the glitches, jumps, temperature drifts and cooler burps. There was only the final step: convert the timelines into a picture of the sky.

We didn't know what we would see. We had all seen the spectacular *Herschel* pictures of interstellar dust (Chapter 9), but we were now trying to do something much more difficult: detect ultra-faint sub-mm radiation from galaxies billions of years in the past. Our main worry was the instability of the bolometers, which was likely to be a particular problem for us because we were making such long observations. The team had developed a very elegant technique for using the thermistor measurements to correct the timelines, but we were worried that if the method did not work perfectly there would be long streaks all over the picture.

When he saw it, Simon said he was just relieved everything had worked. I can't say what I felt because I don't even remember seeing the picture for the first time, December 2009

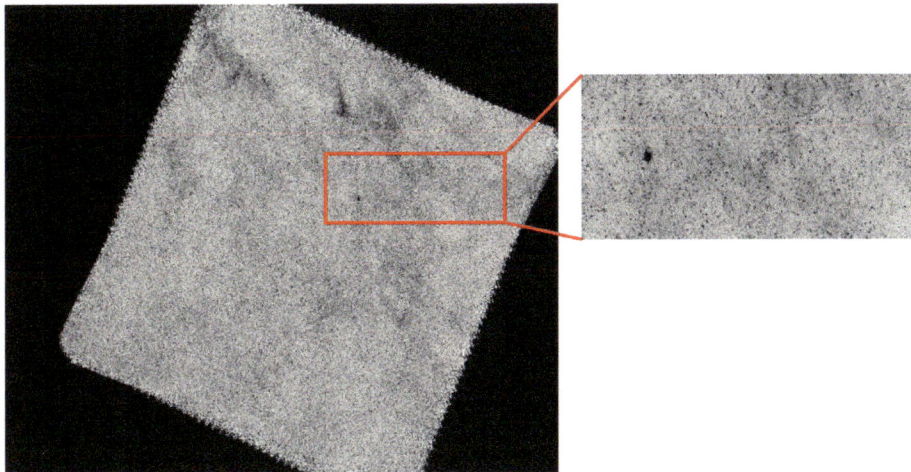

FIGURE 10.2 Our first sub-mm picture. The dark wisps show sub-mm radiation from interstellar dust in our own galaxy. The dark blobs, more discernible in the close-up panel, are sub-mm sources, sub-mm radiation from distant galaxies (the large blob in the close-up panel annoyingly turned out to be a clump of dust in our own galaxy). Although they are very hard to see, there are roughly 7,000 sub-mm sources in this image.

Credit: ESA and the *Herschel* ATLAS

was so busy, with new *Herschel* results every day, Christmas coming, and Göran's conference in Madrid to get ready for.

The picture (Figure 10.2) was the largest ever sub-mm picture of the sky, covering an area of sky roughly sixty thousand times larger than the SCUBA image of the Hubble Deep Field, which only 11 years earlier had been one of the first-ever sub-mm pictures (Chapter 4). Our picture does not have the drama of the *Herschel* pictures of the interstellar dust in our own galaxy (Chapter 9), although there is some of this nearby dust in our picture – the ethereal wisps that seem to float across it. But from our point of view, these wisps were just a distraction. We were interested in what is visible through the wisps: the dark stippling, which is only properly visible in the close-up panel. In the close-up, each stipple emerges as a blob – a sub-mm source. Each of these blobs is the sub-mm radiation from some distant galaxy, and therefore as we looked at the picture on the screen – even if I can't remember it now – we were looking down through the dust in our own galaxy deep into the past.

We were relieved, but we were also aware that it was now less than five months until the deadline Göran had set for the first *Herschel* papers. The crucial next step, without which we wouldn't be able to write any papers at all, was to identify the galaxies emitting the sub-mm radiation. I lost a lot of sleep during the Christmas holidays of 2009 because I wasn't sure this was even possible.

Our problem was that sub-mm pictures don't show much detail (Chapter 4). Unlike the exquisite pictures of galaxies with *Hubble*, a *Herschel* picture would rarely show any detail unless the galaxy was very close – hence the blobs. A consequence of this lack of detail was that it was usually difficult to identify the galaxy emitting the sub-mm radiation.

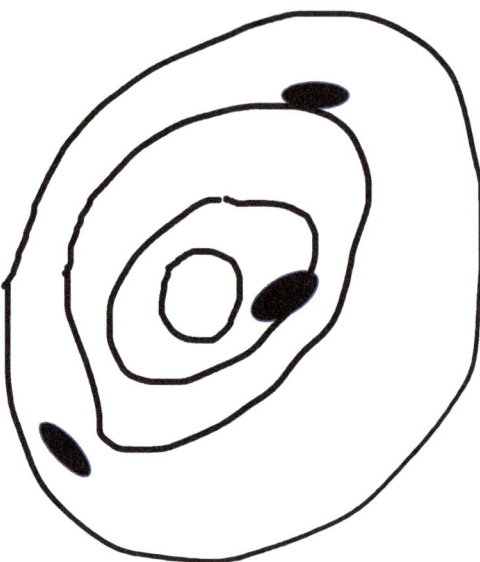

FIGURE 10.3   The problem. The contours show one of the sub-mm blobs, with the contours representing the strength of the sub-mm radiation. The three dark ellipses represent three galaxies in an
optical image that are close enough to the position of the sub-mm source that each of them might
be the galaxy emitting the sub-mm radiation. Which one is it?

The figure I have sketched in Figure 10.3 illustrates the problem. The contours in the
figure show the strength of the sub-mm radiation in the same way that the contours on a
map show height. The little hill of sub-mm radiation is just the same undetailed blob of
sub-mm radiation that appears in images like Figure 10.2, only represented in a different
way. In the optical images, in which there is much more detail, we often found that there
were several galaxies in the vicinity of the sub-mm source that might be the source of
the sub-mm radiation. In my sketch, I have shown three galaxies, the three black ellipses,
within the blob. Each might be emitting the sub-mm radiation. Which is it?

The people who had tried to answer this question were Loretta and Steve and their team
in Nottingham. Before launch, they had developed a method for calculating the probability
that any galaxy found in the optical image was the source of the sub-mm radiation.[5] They
couldn't identify the galaxy emitting the sub-mm radiation for certain, but they could
at least find the one with the highest probability of being its source. It was an ingenious
method, but nobody knew whether it would work in practice.

When we finished our work in Cardiff and were ready to send our sub-mm images to the
Nottingham team, we found we had another problem. We hadn't known when we would
receive our first data, and so normal life had to go on. In December, Loretta and Steve had
two observing runs on Mauna Kea that had nothing to do with our project. They decided

---

[5]   This method took account of how close the galaxy was to the peak of the source (closer – higher probability) and how
bright the galaxy was in the optical waveband (brighter – less likely to fall so close to the sub-mm source by chance).

to use the week between the observing runs as an opportunity for a holiday. They turned off their phones and flew to Kauai, the verdant island at the western end of the island chain.

When they arrived back at the Joint Astronomy Centre on the Big Island, they found all our emails. When she saw our pictures on the screen, Loretta remembers being amazed at the number of sources. In only a decade, we had gone from looking at sub-mm pictures in which there were only a handful of sources (Chapter 4) to ones in which there were many thousands.

Almost immediately, though, they had to drive up the mountain for their observing run. For the next week, they tried to work on the *Herschel* data while on an observing run on Mauna Kea, one of them making the observations while the other sat at the back of the control room working on the *Herschel* data. After the observing run, they flew back to the UK, and then a few days later they drove over to their cottage in the Welsh mountains where they spent Christmas and New Year. While they were there, they often worked on the *Herschel* data until two in the morning. For them, Christmas 2009 never really happened.

While she was still on the summit, Loretta began to look at the galaxies in the optical images close to the positions of the sub-mm sources. Sitting on the top of Mauna Kea, looking at the images on her laptop, she was the first person to see what kinds of galaxies are strong sub-mm sources. Sometimes there was nothing visible in the optical image at the position of the sub-mm source. Sometimes there were several galaxies there and it wasn't clear which was the sub-mm source. But other times there was only a single galaxy close to the position of the sub-mm source and it was obvious what was emitting the sub-mm radiation.

In these cases, she discovered that the galaxy was usually blue. In many of the pictures in this book the colour is not a real colour, but in this case it is. She found that these galaxies were usually a little too faint for her to see their structure clearly, but she thought they generally looked like some sort of spiral or irregular galaxy (Figure 10.4). She wasn't surprised. Spirals and irregulars are galaxies with high stellar birth rates, and they are blue because of the large number of short-lived massive stars (Chapter 4). Stars are formed out of gas; gas and dust are found together; and dust emits sub-mm radiation. It made sense to her that these are the galaxies emitting the sub-mm radiation.

But she also noticed that a few of the galaxies were red (again a real colour). She thought the red galaxies looked like elliptical galaxies (Figure 10.5). This was more surprising. The stellar birth rate in an elliptical galaxy is very low, there is very little gas, so there shouldn't be much dust. They really shouldn't be sub-mm sources.

Even before Loretta and Steve had driven down from the summit, they realised the explanation of these red galaxies.

While Loretta had been staring at the optical images, Steve had been playing with the *Herschel* images. By combining the images we had at different wavelengths, Steve was able to estimate the redshifts of the sources.

I have illustrated his method in the figure (Figure 10.6). The three arrows show the three SPIRE wavelengths: 250, 350 and 500 micrometres. The thick black line shows how the strength of the radiation from the dust in a typical nearby galaxy depends on wavelength. The radiation from a nearby galaxy is usually strongest at about 100 micrometres,

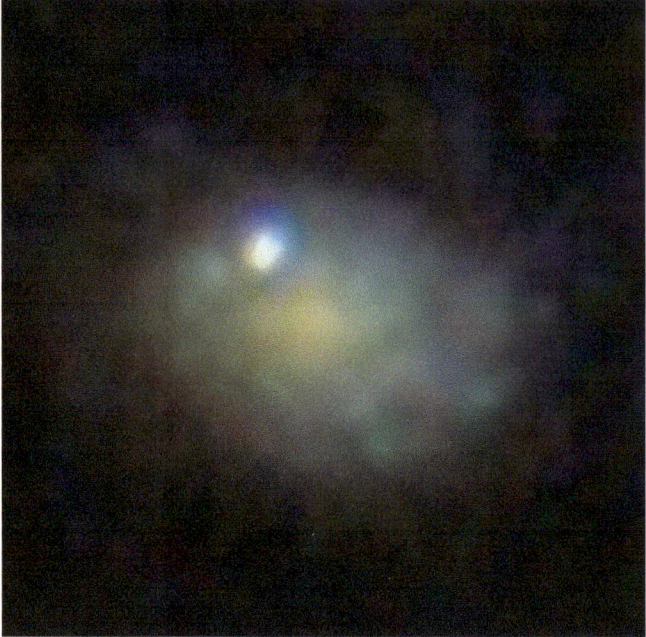

FIGURE 10.4   One of the blue galaxies at the position of one of our sub-mm sources. Most of the galaxies Loretta saw looked like this.

Credit: The *Herschel* ATLAS team

FIGURE 10.5   One of the few red galaxies Loretta saw at the position of one of the sub-mm sources.

Credit: The *Herschel* ATLAS team

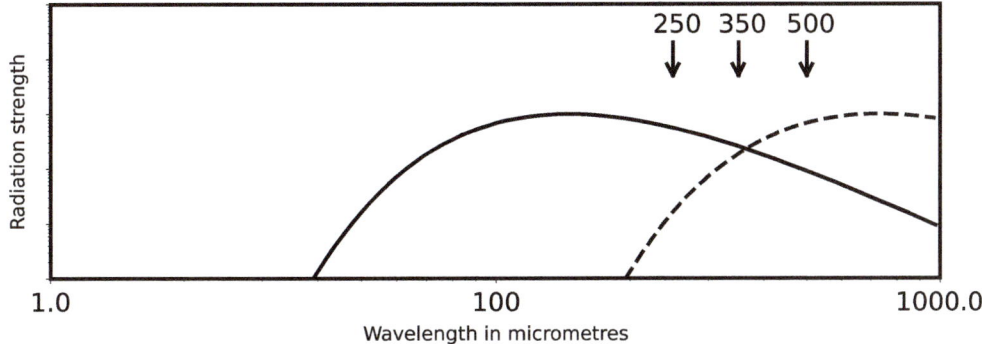

FIGURE 10.6 Our method for estimating the redshift of a source. The solid and dashed lines show how the strength of the radiation from the dust in a galaxy depends on wavelength, for a nearby galaxy (solid line) and for a galaxy at a redshift of 4 (dashed line). The three arrows show the three SPIRE wavelengths: 250, 350 and 500 micrometres. The nearby galaxy is brightest at 250 micrometres. The galaxy at a redshift of four is brightest at 500 micrometres. By comparing the strengths of the radiation from a source at the three SPIRE wavelengths, Steve was able to estimate the redshift of the source.

as it is in the figure, and is less at longer and shorter wavelengths. We therefore knew that if the galaxy emitting the sub-mm radiation was nearby, it was likely to be brightest at 250 micrometres, which is closest to 100 micrometres, and fainter at the two longer SPIRE wavelengths.

But suppose that the galaxy emitting the sub-mm radiation is a long way back in time and has a high redshift. The effect of the redshift is to shift all the radiation to a longer wavelength. In the figure, I have imagined that there is a galaxy with a redshift of 4. The dashed line shows how the strength of the radiation from the dust in a galaxy at a redshift of 4 would depend on wavelength. The galaxy is now brightest at the longest of the SPIRE wavelengths, 500 micrometres, and faintest at the two shorter wavelengths.

And so it is possible to estimate the redshift of a source from the strength of its radiation at the three wavelengths. If the source is brightest at 250 micrometres, fainter at 350 micrometres and faintest at 500 micrometres, the galaxy emitting the sub-mm radiation is likely to be nearby. If it is brightest at 500 micrometres, fainter at 350 micrometres and faintest at 250 micrometres, it is likely to be at a very high redshift.

Steve combined the SPIRE images at 250, 350 and 500 micrometres into a single picture, using colour to represent the probable redshift of the sub-mm source estimated using this method (Figure 10.7). He chose red (now not a real colour) to represent a source at a very high redshift.

The first thing Loretta and Steve noticed when they looked at this picture was how many red sources there were. When Loretta and I had the original idea for the survey, we thought that it would be a great way of finding interesting nearby galaxies. When we seriously started to plan the survey at the meeting at ESTEC, Asantha Cooray told us we were being too unambitious (Chapter 6). He said that SPIRE would be so sensitive that we would also find many of the high-redshift sub-mm galaxies, like the ones discovered with SCUBA. Steve's picture showed that Asantha had been right.

FIGURE 10.7   Steve's picture.[2] The colours (in this case not real colours) have been chosen to show the probable redshifts of the sources. Red shows a source that is probably at high redshift, blue one at low redshift.

Credit: ESA and the *Herschel* ATLAS team

They also noticed something else. Whenever Loretta found a single red galaxy (real colour) in the optical image near one of the sources, they found that in Steve's picture the source was also red (not a real colour[**]), showing that it was at a high redshift. They both realised immediately that these red (real colour) galaxies must be the gravitational lenses that Gianfranco de Zotti had claimed we would find (Chapter 6). The red elliptical galaxies in the optical images were nearby galaxies and not emitting sub-mm radiation. The true galaxies emitting sub-mm radiation were much further away, too far to see in the optical images, with their sub-mm radiation magnified by gravitational lenses, the red galaxies in the optical images, in front of them (Chapter 11).

Gianfranco had been right as well.

For most of our sub-mm sources, though, the identity of the galaxy emitting the sub-mm radiation was not so obvious. Like the source in Figure 10.3, it was not clear which of several galaxies was the source of the sub-mm radiation. Loretta and Steve's team set to work calculating probabilities.

[**]   I know this is confusing, but when we were choosing a colour scheme to represent redshift, I never thought I might have to describe it in a book someday.

By the time I went back to work in January 2010, the Nottingham team had finished their analysis. We discovered about 7,000 sub-mm sources in our first images. The Nottingham team managed to identify the galaxy probably emitting the sub-mm radiation for about one-third of them. Because of the spectroscopic survey of this field (Chapter 6), we already knew the redshifts of most of these galaxies. The highest redshift was about 0.5, much lower than the redshifts of the sub-mm galaxies discovered with SCUBA (Chapter 4). We realised that the other two-thirds of the sources must be at higher redshifts, like the sub-mm galaxies discovered with SCUBA, so that the galaxies emitting the sub-mm radiation were too faint to see in our optical images.

We decided to write some papers.

The deadline for these was 8.00 pm on 13 May. Göran had made a deal with the editor of *Astronomy and Astrophysics*, the main European astronomy journal, that there would be a special issue for the first papers from *Herschel*. The H-ATLAS team as a whole wrote eight papers that appeared in the special issue. In Cardiff we wrote one.

Since we knew the redshifts for most of the galaxies identified by the Nottingham crew, we were able to calculate the sub-mm luminosity of each galaxy – how much sub-mm radiation the galaxy is actually emitting rather than the strength of its sub-mm radiation as measured by *Herschel* (which depended on both the luminosity of the galaxy and how far away it is). We decided to write a paper about almost the simplest thing one can derive from a survey like ours: the 'luminosity function'. Despite the grand name, this is just a set of estimates of how many galaxies there are in a fixed volume of space in different luminosity ranges. In the same way that there are always more small lakes than large lakes

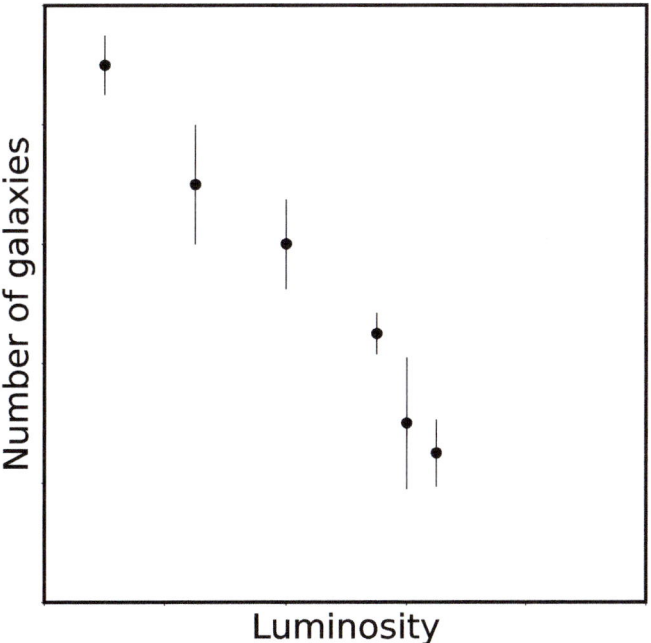

FIGURE 10.8   A typical luminosity function.

and more small hills than high mountains, there are always more galaxies that have a low luminosity than ones with a high luminosity, so most luminosity functions look a little like the imaginary one I have sketched in Figure 10.8. Even though it's imaginary, I have given it some error bars to illustrate that there is always uncertainty in any measurement.

We realised that we could do better than calculate a single luminosity function because we had estimates of the luminosity for so many galaxies. A redshift of 0.5 doesn't sound very high compared to the redshifts I have mentioned elsewhere in this book – the redshifts of about 3 for the sub-mm galaxies or the redshifts of over 10 now being routinely measured with the *James Webb Space Telescope* – but it's still a long way back in time, about five billion years (500 million years before the Sun and all the planets were even formed).

We decided to split the galaxies into five separate redshift ranges and calculate a separate luminosity function for each range. We chose redshifts from 0 to 0.1, 0.1 to 0.2, 0.2 to 0.3, 0.3 to 0.4 and 0.4 to 0.5 – each redshift range containing galaxies about one billion years further back in time than the one before, back to about five billion years before the present day. By comparing the luminosity functions in the different redshift ranges, we would be able to look for changes in the luminosity function over the last five billion years – evolution.

I didn't expect to see any. Ten billion years ago, the universe was a very different place. The discovery with SCUBA of the sub-mm galaxies, the ancestors of present-day ellipticals, had already showed us this. And there's plenty of other evidence, including the changes in galaxy sizes (Chapter 6), for big changes between the universe ten billion years ago and the universe today. But compared to the age of the universe (14 billion years), five billion years didn't seem very large. I didn't expect there to have been much change in this time.

We had decided that Simon Dye, the postdoc who had done much of the hard work making the sub-mm images, should be the one to do the analysis and write the paper. I was writing papers about two other *Herschel* projects, so for the next few months Simon and I had little chance to talk. Then one day he phoned me and asked me to come upstairs to his office.

When I got there, he showed me his results. He showed me on his computer screen a figure showing the luminosity function for each redshift range, each shown in a different colour (Figure 10.9). I had expected that they would all be roughly the same, but they weren't. The luminosity function for each redshift range was displaced to the right of the one before, shifted to a higher luminosity. His figure showed that galaxies in the past emitted more sub-mm radiation than galaxies today. We already knew that the sub-mm galaxies discovered with SCUBA, ten billion years in the past, emitted a lot more sub-mm radiation than galaxies today, but the luminosity functions that Simon had plotted were for galaxies only a few billion years in the past. On his screen, I could see that even one billion years ago (the luminosity function shown in red), the galaxies were emitting more sub-mm radiation than they do today (the luminosity function shown by the black line). And in cosmic terms, one billion years is only yesterday. We were the first astronomers to see evidence for such recent evolution in the galaxy population.[3]

Deadline day. It is a beautiful spring evening. I am at home about to push the key on my laptop that will submit my paper. Simon is at home in Bristol about to submit his paper.

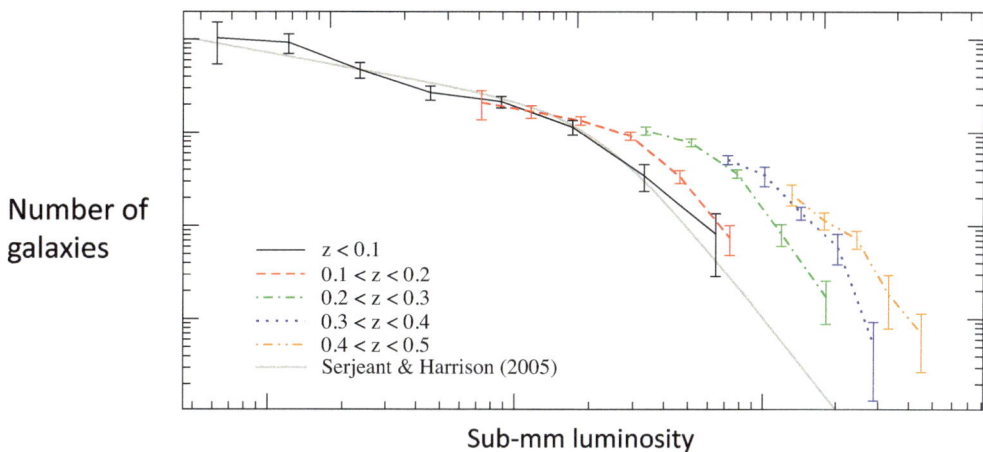

FIGURE 10.9 The sub-mm luminosity function in five redshift ranges, each shown in a different colour: black – redshifts 0.0 to 0.1; red – redshifts 0.1 to 0.2; green – redshifts 0.2 to 0.3; blue – redshifts 0.3 to 0.4; beige – redshifts 0.4 to 0.5.

Credit: Simon Dye and the *Herschel* ATLAS team

When I push the key, my paper will be sent to the internet astronomy archive. Thirty-six hours later, astronomers will be able to read it with their morning coffee.

It's the beginning of something, it's the end of something. For months I have been thinking about my paper – at work, in the shower, while watching TV – and now it's all over. I am about to send my paper out into the world. I don't have any illusions that it will have much impact, I know it's not that exciting. I think Simon's paper is one of the more important ones, but you never know how the astronomy community will react.

CHAPTER 11

The Birth of a Galaxy

BY THIS TIME, WE had already made two mistakes.

Loretta and I were having a drink in a bar when we realised the first one. We were in Noordwijk, a small seaside town I like to stay in when I visit ESTEC. We were at the conference that Göran organised in May 2010 and were at one of the bars on the seafront, depressurising after a long day and watching the sun go down over the North Sea.

Planning our survey, the biggest that would be carried out with *Herschel*, had been a challenge. It would only be possible to observe our two big fields during limited visibility windows and we had not been sure whether there were enough of these during the three-year mission to finish the survey. Loretta and Steve Maddox were the ones who had taken on the challenge. Using the *Herschel* scheduling software, they had succeeded – just – in fitting all the necessary observations into the available visibility windows. By the time the helium ran out, we would be able to make two huge images, in opposite directions, one close to the north Galactic pole* and one close to the south Galactic pole, the largest sub-mm images of the sky ever made.

But going back over what had happened during the day, we realised there was a problem. We were planning to use a special observing mode in which both cameras were used at the same time. We had learned during the day that only a fixed number of days had been allocated to each observing mode. While talking in the bar, we both realised at almost the same moment that Loretta and Steve had not taken this into account when they had planned the observations. And then we suddenly realised there was another problem. Something else we had learned during the day was that the mission would be divided into operational days (ODs), with 21 hours for observations and a three-hour contact period that would be used to talk to the telescope and download data. Loretta and Steve hadn't realised this either. In their plan, some of our observations would last 15 hours, which would waste the last six hours of the observation period. We went to bed that night very worried.

We were still worried when we went into ESTEC the next morning. At the first coffee break, we found Göran and explained the problem. He summoned some of the backroom

---

* Where these are is explained in Chapter 6.

DOI: 10.1201/9781003195290-11

ESA technical wizards. While the rest of the conference went on, we had a long meeting in a side room with the sound of the talks just audible through the wall. After much discussion, the wizards thought we might still be able to complete our survey if we redesigned it so that each observation lasted 10.5 hours, which would mean that two observations would fit neatly into each OD.

Calmed down a little, we managed to sit through the rest of the conference. Back in Nottingham, Loretta and Steve started to redesign our observing plan. When they had finished, they had succeeded in fitting in all the necessary observations. We would be able to complete the survey before the helium ran out – just.

The second mistake was less forgivable.

We had forgotten to look properly, which may seem rather strange since that is what observers like us do for a living. The difficulty is that you see what you expect to see. When Loretta sat on the summit of Mauna Kea, flipping through pictures of galaxies on her laptop, she thought that most of the galaxies she saw at the positions of the sub-mm sources looked like standard spirals and irregulars. We had all seen these pictures. We had all thought the same thing: that the galaxies were the same spirals and irregulars that astronomers have been finding in surveys in the visible waveband for years.

One useful way to classify galaxies is to plot them on a graph of optical luminosity (how much optical or visible light is emitted by the galaxy rather than how much we detect on Earth) versus colour (Figure 11.1). The galaxies discovered by surveys in the visible waveband mostly lie in two areas of this diagram: a strip on the red side, which astronomers have named the 'red sequence', and a less clearly defined region on the blue side, which we call the 'blue cloud'. There are fewer galaxies in the in-between region, which astronomers call the 'green valley'. The galaxies in the blue cloud, which are mostly spiral and irregular galaxies, are ones in which there is a high stellar birth rate, and the large number of short-lived massive blue stars in these galaxies explains their blue colours. In the galaxies in the red sequence, which are mostly ellipticals, the stellar birth rate is very low, there are

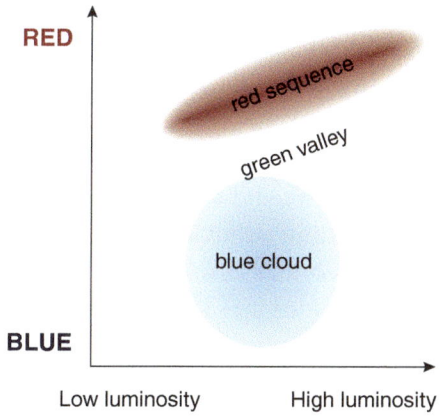

FIGURE 11.1 Colour versus optical luminosity (the total amount of light emitted by a galaxy rather than the amount detected on the Earth) for the galaxies found in a survey in the visible waveband.

very few of the short-lived massive blue stars, and all that is left are the low-mass stars with long lives, which are red (Chapter 4).

Before we started our sub-mm survey, almost everything we knew about galaxies came from these surveys in the visible waveband, which have been carried out since the time of the Herschels. We didn't expect to detect in our sub-mm survey any galaxies on the red sequence, which contain very little gas and therefore very little dust, the stuff that is necessary for sub-mm radiation. This is why Loretta was surprised to see in the optical images red galaxies at the positions of some of our sources, and why she quickly realised they must be gravitational lenses (Chapter 10). We *did* expect to detect galaxies in the blue cloud because they contain a large amount of gas and dust. So when we saw blue galaxies at the positions of some of our sub-mm sources (Chapter 10), we thought we were seeing the same spirals and irregulars found in surveys in the visible waveband. We never noticed there was anything different about them.

We saw what we expected to see.

It is a little ironic. The original motive for our survey was that Loretta and I had wanted to know whether there were any differences between the galaxies found in a sub-mm survey and those found in surveys in the visible waveband. We had somehow now forgotten this.

It took eight years for us to realise our mistake, and that we had missed an important discovery (Chapter 15). Our only excuse was that we were distracted by another discovery. This discovery allowed us, for the first time, to watch the birth of a galaxy.

The ideal gravitational lens is a massive galaxy or even better a cluster of galaxies. When one of these happens to fall in front of a more distant object, the gravity of the lens bends the light (or any other type of radiation) from the more distant object (Figure 11.2). The light from the more distant object is now arriving on Earth from several directions, which means that an observer, looking back along the direction of the light, sees more than one image of the object. In the figure I've shown the light arriving from just two directions, but it may arrive from many directions at once.

Anything viewed through a gravitational lens is easy to spot. The clue is, it doesn't look natural.

Most galaxies look natural in the same way a tree looks natural, each one slightly different from every other. A galaxy looks like it has grown organically. It is not surprising that when William Herschel started finding nebulae in his 'nebulous beds', the first words he reached for were from the natural world.

Anything viewed through a gravitational lens, on the other hand…well, just look for yourself. Figure 11.3 shows some examples I have scooped from the internet. The swooping arcs around the cluster of galaxies in the top-left panel make it look as if the cluster is being viewed through a portal in space, something out of Star Trek perhaps. The lights in the top-right panel – a Catherine wheel about to spin? – seem too precisely positioned to

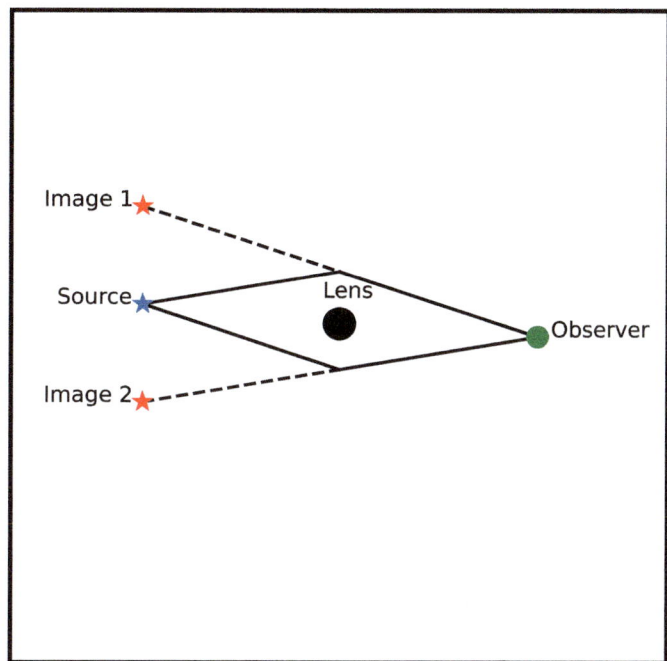

FIGURE 11.2   The effect of a gravitational lens. The light travels from the source to the observer and is bent by the gravity of the lens, a massive galaxy or a cluster. As a result, the light arrives at the observer from multiple directions (two in this figure). Looking back along the direction of the light, the observer sees multiple images of the source.

be there by chance. And the rings around the galaxies in the bottom panel are just weird. Is someone trying to tell us something?

These bizarre images are the result of the light arriving at the observer from several directions, so that an observer, looking back along the direction of the light rays, sees multiple images of the source. In the top-right panel of Figure 11.3, the lens is the orange object in the centre. The light from a quasar behind the lens is bent by the lens so that it arrives from four different directions, which means we see four separate images of the quasar. Sometimes the light arrives from so many directions that the images merge to form an arc, as has happened around the cluster in the top-left panel, or even into a perfect ring if the object, observer and lens are precisely aligned, as is happening in the bottom panel. (Ironically, these images that look so artificial are the result of a simple natural process.)

A magnifying glass in which an object appears as multiple images or an arc or a ring wouldn't be much use, but the mathematics of the distortion is well understood and it's possible – if you know how to do it (or know the right person) – to correct for the distortion. Once the distortion has been removed, the image contains more detail than if the lens had not been there, which is one of the reasons astronomers are so excited about them. They are huge cosmic magnifying glasses.

Gravitational lenses have captivated astronomers since their discovery almost 50 years ago.[1] But astronomers have never been quite sure what to use them for. Apart from the obvious idea of using the lens as a magnifying glass, there have been many other ideas

FIGURE 11.3   Some examples of the effect of a gravitational lens. In the top-right panel, the light from a quasar, after being bent around the lens in the centre, arrives from four different directions, producing four separate images of the quasar. In the other panels, the light from the object behind the lens arrives from so many directions that the multiple images merge into an arc or a ring.

Credit: NASA/ESA & A. Bolton and the SLACS team, J. Rigby, K. Sharon, M. Gladders and E. Wuyts & S.H. Suyu et al.

over the years for how we might use them as a tool, one of the most interesting being to use a large number of lenses to measure the amounts of dark matter and dark energy in the universe.[†]

The big problem with this idea, though, is that although it is fairly easy enough to spot something that is lensed, nobody had come up with a good way of finding a large number of them. It's probably fair to say that at the time of the launch of *Herschel*, gravitational lenses had not measured up to all the hype.

[†]   See my book *Origins* for more about dark matter and dark energy.

The birthplace of Mattia Negrello is a long way from the shiny, high-tech world of *Herschel*. A small town in north-eastern Italy, Rovigo is only about 40 kilometres from Padua, where Galileo taught, but it might as well be on the other side of the Moon. Rovigo is deep in the Italian countryside, everyone speaks one of the Italian dialects, incomprehensible to anyone outside the local area, and education is limited – Mattia's parents left full-time education after primary school.

It was while he was carol singing that Mattia became interested in astronomy. On winter evenings, away from the city lights, he became enthralled by the night sky. One January he started saving money. His parents assumed he was saving up to buy a bicycle, which is what all the other boys were doing at the time. Instead, after a few months he went out and bought a telescope. His parents, like everyone around them, knew next to nothing about the universe, not even the difference between the stars and the planets. After Mattia showed him the Moon and Jupiter through his telescope, his father was not able to sleep that night. The following morning, he thanked Mattia. He said he had not been able to stop thinking about what he had seen. He had never had such an experience before.

For a long time, Mattia was undecided between art and science, but eventually, inspired by two passionate high-school teachers of physics and maths, Elisabetta Lorenzetti and Sandra Berneccoli, he travelled the 40 kilometres from Rovigo to Padua, where he started a degree in physics. Once he finished this degree, he travelled a little further around the bend of the Adriatic to Trieste, where he enrolled as a PhD student at the *Scuola Internazionale Superiore di Studi Avanzati* (SISSA), which is one of the top places in the world to do research in astrophysics.

His PhD supervisor was Gianfranco de Zotti. Gianfranco is a soft-spoken, very courteous astronomer, who unlike the rest of us always wears a suit. Despite his old-fashioned manners and appearance, he is always prepared to question the conventional view. Around the time Mattia moved to SISSA, Gianfranco had just spent ten years running his department. He was keen to get back to research. He had become interested in how a galaxy is born. Around this time, most astronomers had come to believe that the very idea of a 'birth' is misleading, deciding that the formation of a galaxy occurs over a long period rather than the galaxy being born at a particular time. Of course, the formation of a galaxy, a massive object, does not happen in an instant, but most astronomers working in the field had reached the extreme conclusion that it effectively takes the entire age of the universe. In the orthodox view at the time, a galaxy is formed by a series of mergers of smaller galaxies, with each merger triggering the formation of more stars out of the gas in the merging galaxies. Since these mergers are still happening today, the formation of a galaxy never really stops.

Gianfranco thought they were wrong. He had been struck by the discovery of the sub-mm galaxies with SCUBA (Chapter 4). He realised that the stellar birth rates in these galaxies were so high that it would require only about 100 million years to form enough stars to make a giant galaxy. Since this was only about one per cent of the age of the universe (not too different, as it happens, from the percentage of our life we spend in the womb), he thought that for these galaxies, at least, the term 'birth' seemed about right. He decided to create a computer model based on the idea that an elliptical galaxy is formed very quickly – is born – as the result of the gravitational collapse of a large cloud of gas.

Around this time, Mattia became his PhD student. Gianfranco needed to find him a project. He had recently heard about a new sub-mm telescope that would soon be launched by ESA. He asked Mattia to use the model to predict how many sub-mm galaxies, which he was convinced were galaxies being seen during their birth, would be detected by the new telescope.

When Mattia came back, he told Gianfranco that he had discovered something very interesting. He said that if the new space telescope was used to carry out a sub-mm survey of a very large area of sky, many of the brightest sources found by the survey, according to the model, would be behind gravitational lenses.[2] Gianfranco realised that a survey like this would be an excellent way of finding a large number of gravitational lenses. To make sure such a survey happened, he started to go to the meetings that had been arranged to plan the observing programme for the new telescope. Loretta and I met him at the meeting at ESTEC in 2007 (Chapter 6). A part of the proposal we wrote was based on Gianfranco's claim that a survey of a large area of sky would be a great way to find gravitational lenses. I didn't really believe it – theorists have a lot of ideas – but it made a nice section in the proposal.

Everything changed when Loretta started to spot red (real colour) galaxies at the positions of our sub-mm sources. She and Steve quickly realised these must be Gianfranco's gravitational lenses (Chapter 10).

One of their best candidates was the source SDP 81.[‡] In the optical image there was a red galaxy close to the source (Figure 11.4), which Loretta thought looked like an elliptical galaxy, a type of galaxy that doesn't usually emit sub-mm radiation. The galaxy had a redshift of 0.299. As I described in the last chapter, Steve used the relative strength of the sub-mm radiation at the three SPIRE wavelengths to estimate the redshifts of our sources. His estimate was that the redshift of SDP 81 was about 3, much higher than the redshift of the galaxy in the optical image. Loretta and Steve decided that the red galaxy itself was probably not the one emitting the sub-mm radiation, but was instead a lens, magnifying the sub-mm radiation from a more distant galaxy, which they could not see in the optical image because it was too faint. They found several sources like this.

It all hung together. But we knew nobody outside our team would believe us without some stronger evidence.

By this time, Mattia had travelled a long way from Rovigo. After finishing his PhD in Trieste, he moved to Milton Keynes in the UK, where he worked as a postdoc at the Open University with Stephen Serjeant, a professor who had been one of the founding members of our survey (Chapter 6). Mattia took on the challenge of proving the red galaxies were really lenses. At the time, even some members of our team thought this was a stupid idea. One senior member, a professor at a top UK university, even told Mattia he was ruining his career working on the subject.

What we really needed was a sub-mm picture with enough detail to show that the sub-mm radiation was being bent by the lens. Multiple images would be great, arcs, like the ones in Figure 11.3, even better. But the only sub-mm picture we had so far was the

---

[‡] We called it SDP 81 because it was the 81st source in the catalogue of sub-mm sources we discovered during the Science Demonstration Phase (SDP). It has an official name, but our name is still used.

FIGURE 11.4 The red (real colour) galaxy seen in a visible image at the position of the sub-mm source SDP 81.

Credit: The *Herschel* ATLAS team

Herschel one that showed SDP 81 as the usual blob. One way to see more detail is to build a telescope with a bigger mirror (Chapter 4), but in this case I reckoned that we would need a mirror 100 metres across to take a sub-mm picture that showed enough detail[§].

Fortunately, there was another way.

The other way of seeing more detail is to use an interferometer, an array of smaller telescopes in which a picture is produced by combining the signals from the individual telescopes. The level of detail depends on the distance between the telescopes. If the array of telescopes is spread over 100 metres, for example, the pictures taken with it will show as much detail as one taken with a single telescope with a mirror 100 metres across – and an interferometer is a lot cheaper and easier to build.[¶]

And luckily for us, there was now one interferometer we could use to make observations in the sub-mm waveband. The SubMillimeter Array or SMA (Figure 11.5) is an array of eight small telescopes, located just below the summit of Mauna Kea, which uses one of the narrow atmospheric windows in the sub-mm waveband (Chapter 2). The window is a little longward of the wavelengths we had been using with *Herschel*, but for our purpose this didn't matter.

[§] As I explained in Chapter 4, one of the big problems with sub-mm astronomy is that you need an extremely big telescope to take pictures that show the same level of detail as pictures taken with a small optical telescope.

[¶] There is a downside. The interferometer is not so sensitive because the combined collecting area of all the small telescopes is not as large as of a single telescope with a 100-metre mirror.

FIGURE 11.5   The SubMillimeter Array or SMA.

Credit: the Smithsonian Astrophysical Observatory

For most telescopes, the deadlines are every six months or every year, which is frustrating if you want some observations quickly, but there is also the Director's Discretionary Time. This is the small amount of observing time reserved for the telescope director to allocate for very urgent projects – a supernova that has just gone off, etc. We couldn't really argue our project was urgent, but we did think our project was exciting and we did urgently want to know the answer. We hoped the director might agree. We had five strong lens candidates. Mattia picked our best one, SDP 81, and submitted a proposal.

The director did agree. Only a few weeks later, the SMA started observing SDP 81. When the observations finished, Mark Gurwell, our SMA support scientist, reduced the data.

When he displayed the finished image on the screen, however, he couldn't see anything there. He asked Mattia to check he had given him the correct position. The position had been correct, but it was still a little embarrassing. Directors are usually happy to give you observing time for an exciting project, but they do expect you to detect something.

Mattia could think of only one possible explanation. The telescopes in an interferometer are not fixed in position and can be moved into different configurations. If you want to see the greatest possible detail in the picture, you ask for a configuration in which the telescopes are as far apart as possible, but the broad-brush structure will then not show up so well. If you want to see the broad-brush structure and don't care so much about detail, you ask for a configuration in which the telescopes are as close together as possible. Mattia wondered whether he had asked for the wrong configuration. He decided to ask for some more observing time with a different configuration. The SMA Director doled out some more observing time. The SMA made the observations. Mark Gurwell reduced the data again. He emailed the image to Mattia.

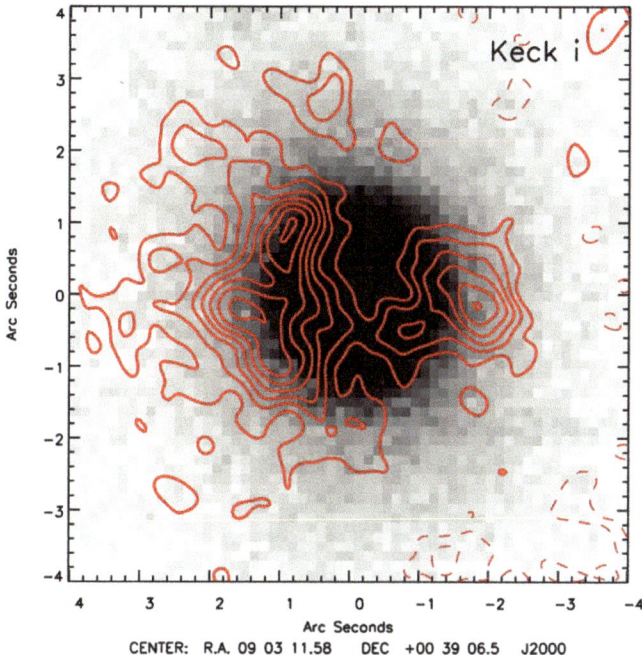

FIGURE 11.6 Picture of SDP 81 that shows both the visible light and the sub-mm radiation. The visible image, in shades of grey, shows the galaxy that is the gravitational lens. The sub-mm radiation, represented by the red contours, is from a galaxy that is too far away to be seen in the visible image. This galaxy's strong sub-mm radiation has been bent by the gravitational field of the lens so that the red contours seem to wrap around the lensing galaxy – arcs!

Credit: The *Herschel* ATLAS team

By now we had also a much better picture in the visible waveband. Asantha Cooray, one of the other founders of the survey, had access to the Keck Telescope, the largest optical telescope in the world, which he used to take a picture of the red galaxy (real colour) at the position of SDP 81. Mattia combined the Keck and SMA images into a single picture showing both the visible and sub-mm radiation. He immediately went round to show it to his boss, Stephen Serjeant. A little later we all saw it.

As we often do (Chapter 3), Mattia represented the visible and sub-mm radiation in a different way in the picture (Figure 11.6). Ignoring the real red colour of the galaxy, he showed the strength of its visible light in shades of grey. He represented the strength of the sub-mm radiation by red contours[**], a colour he chose, not in this case because it had any significance, but because it showed up better against the grey. The first thing we all noticed when we saw the picture was that the contours showing the sub-mm radiation did not peak in the same place as the visible light. They skirted around the edge of the galaxy, almost embracing it.

[**] I realise I have now used red for three different things. Sorry!

Arcs!

As evidence this was a lensed system, the arcs were quite contvincing. But, to be absolutely sure, we also wanted a definite redshift – not just Steve's estimate – for the galaxy emitting the sub-mm radiation. If we could show this was higher than the redshift (0.299) of the red galaxy in the visible image, that would be enough to nail it. We needed a spectrum.

Optical spectroscopy was out. We couldn't even see the galaxy emitting the sub-mm radiation in the visible image, it was so faint. But now we were very lucky. There are many bright spectral lines in the sub-mm waveband, mostly from common molecules like water and carbon monoxide, but until very recently – the technology in our waveband is so far behind everyone else's[††] – there were no spectrometers we could use to observe these lines. Fortunately, just at the moment we needed one, there were now experimental spectrometers on a few sub-mm telescopes.

While Mattia had been working on the SMA data, some of the others in the team had been sending proposals to the directors of the three telescopes that had one of these new spectrometers: the Caltech Submillimeter Observatory on Mauna Kea, the Greenbank Telescope in the Appalachian Mountains and the Plateau de Bure telescope in the French Alps. On 9 March 2010, only three months after we had seen our first *Herschel* picture on the screen, we were sent a sub-mm spectrum of SDP 81 that had been taken with the telescope on Mauna Kea. Two weeks after that we received one taken with the telescope in the French Alps. A few days later, we received one taken with the telescope in Appalachia. They all showed the same thing: spectral lines (Figure 11.7).

The precise redshift of SDP81 is 3.04. The redshift of the red elliptical galaxy in the visible image at the source's position is 0.299, proving that this galaxy is not the one emitting the sub-mm radiation but is a gravitational lens. The sub-mm radiation is from a more distant galaxy (invisible in the optical picture), which is being bent by the gravitational field of the lens.

With some more observations, we were soon able to show that all five of the sources on Mattia's list are behind a gravitational lens. With five gravitational lenses discovered from our very first *Herschel* image, we estimated that several hundred would be discovered in the full survey. Mattia wrote a paper, which he published in the journal *Science*, announcing a new way of finding gravitational lenses.[3]

The best thing about these new gravitational lenses were the objects behind them. The sub-mm galaxies discovered in the SCUBA surveys (Chapter 4) had always been tantalising objects. Ten billion years back in time, with stellar birth rates high enough to make all the stars in a giant galaxy in a blink of a cosmic eye, I and everyone else who carried out the surveys with SCUBA had always thought these were galaxies being seen during their formation phase – galaxies being born. For the last 15 years, we had tried to learn more about them, but all the dust made them so hard to study with conventional optical telescopes that we hadn't learned very much. They had stayed sub-mm blobs. Now we had

---

[††] Optical astronomers have used spectrometers since the 1860s when William and Margaret Huggins used one to show that the stars are made from the same chemical elements found on earth.

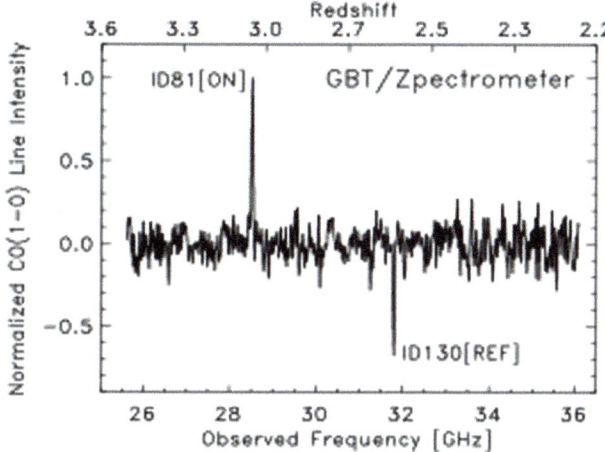

FIGURE 11.7   Our spectrum of SDP 81 obtained with the Greenbank Telescope in Appalachia. The upwards spike shows a spectral line from carbon monoxide (the downwards spike is a spectral line from a different galaxy).[3]

Credit: The *Herschel* ATLAS team

discovered some behind gravitational lenses, huge cosmic magnifying glasses. We had the chance to study them in detail.

We had the chance to see a galaxy being born.

While we had been busy with *Herschel*, a giant telescope had been taking shape.

With the Atacama Large Millimeter Array, it's easy to run out of superlatives. An interferometer linking 64 individual dishes in the Atacama Desert in the Andes (Figure 11.8), ALMA is the highest telescope in the world, eight hundred metres higher even than the telescopes on Mauna Kea. With a total cost of 1.4 billion dollars, it is the most expensive telescope ever built on the ground. And it is the most over-subscribed telescope, on earth or in space, with only about 1 out of every 10 observing proposals being accepted.

Best of all for us, it is a sub-mm telescope. Its pictures show even more detail than those taken with *Hubble* and the *James Webb Space Telescope*. When ALMA started operations in 2013 sub-mm astronomers finally got the chance to take decent pictures. Goodbye, blobs!

Of course, we wanted to use it to take pictures of the sub-mm sources we had discovered behind the lenses. Even without the lenses, we knew ALMA would give us amazing pictures, but with the magnification from the lenses as well, we knew we would get pictures with unprecedented detail.

Fortunately, this time we did not have to bother with observing proposals and time allocation committees. SDP 81 was chosen by the ALMA staff as the perfect object to test

FIGURE 11.8   The Atacama Large Millimetre Array – ALMA.

Credit: Clem and Adri Bacri-Normier – ALMA (ESO/NAOJ/NRAO)

one of the array configurations. During October 2014, almost four years after we had discovered it in our first *Herschel* image, ALMA observed SDP 81.

Cat Vlahakis, who was part of the ALMA team that made the observations, said she had tears in her eyes when she saw the picture of SDP 81 (Figure 11.9). The almost perfect ring doesn't leave any doubt that the source is being viewed through a lens. At the centre of the ring is the gravitational lens, which we don't see in this sub-mm picture because elliptical galaxies emit hardly any sub-mm radiation. I think the picture is beautiful too, but, with its tendrils fading off into the dark, I also find it slightly sinister. I wasn't the only one who was reminded of the ring in *Lord of the Rings*.

To use the lens as a giant cosmic magnifying glass, we needed to correct the distortion. Fortunately, we knew just the person to do this. Simon Dye had been my postdoc in Cardiff and was now a lecturer in Nottingham. He also happens to be an expert in correcting the distortion from a gravitational lens. When we contacted him, he was at a conference in Birmingham. He thought this was much more important than the conference, so he went to the back of the conference room and set to work on his laptop.

Figure 11.10 shows Simon's reconstruction of the undistorted sub-mm radiation from SDP 81, in which he has used colour to show the strength of the radiation. When I saw it, I found it difficult to believe the detail. The smallest thing in the picture is about 100 light-years across, which is only about the diameter of one of the giant molecular

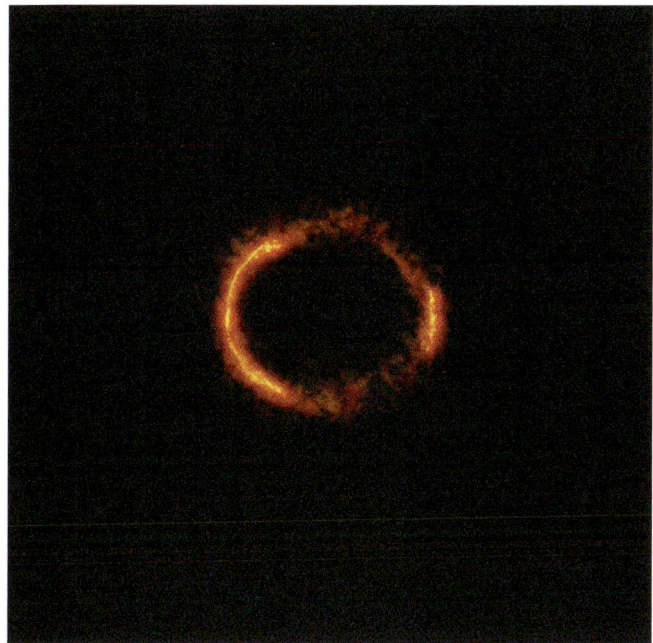

FIGURE 11.9   The ALMA image of SDP 81.[4]

Credit: B. Saxton – ALMA (ESO/NAOJ/NRAO)

clouds in our galaxy (Chapter 9). We are looking back in time 11.6 billion years, but we are still able to see things this small.[5]

This is what a galaxy looks like as it is being born. Remember, we are *seeing* it. Although the birth was 11.6 billion years ago, we are seeing it, thanks to the finite speed of light, as it happens.

When we looked at Simon's picture, we could see, from the blobs in the picture, that at the time of its birth, this galaxy contained a few giant clouds of gas and dust. These were huge compared to the clouds in our own galaxy, with masses roughly 100 times greater.[5] We could also see that at its birth the galaxy was very small. The distance from top to bottom of the galaxy is about 6,000 light-years, compared to the 30,000 light years from the centre of our galaxy to the Sun (and the Sun is not on the edge). Galaxies are very small when they are born.

And they are also very lively. We had already guessed (Chapter 4) that during the birth of an elliptical galaxy, the stellar birth rate must have been exceptionally high, and so the galaxy must have contained a very large number of massive short-lived stars. In SDP 81, we can't see the visible light from these stars because of the dust, but we were able to estimate, from the sub-mm radiation emitted by the dust, that, the stellar birth rate was roughly 160 times greater than in our galaxy today.[6]

The high stellar birth rate in such a small volume shows that the birth was not a tranquil one. The radiation from the massive stars would have irradiated the gas in the galaxy, heating it and ionising it. The gas must also have been swept by shock waves when the stars

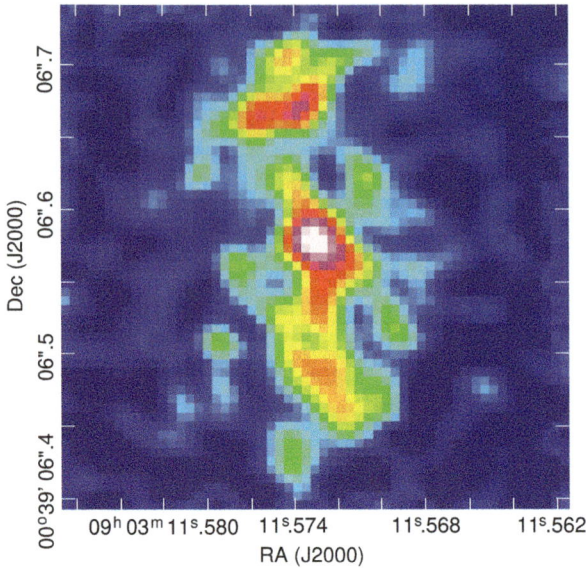

FIGURE 11.10   The undistorted sub-mm picture of SDP 81,[6] with colour used to show the strength of the sub-mm radiation (white, red, green, blue – in order from strong to weak). The blobs in the picture show that during its birth, the galaxy contained huge clouds of gas and dust with masses about 100 times greater than the giant molecular clouds in our own galaxy.

exploded as supernovae. The same activity occurs in spiral galaxies today but not with the same vigour. In spiral galaxies today, the massive stars are spread out along the spiral arms and are not so hidden by dust, which is why the *Hubble* pictures of spirals are so dramatic. In SDP 81, all these massive stars, 160 times more than in our own galaxy, were crammed into a volume roughly 1,000 times smaller.

We were able to learn more about the birth from some other observations of SDP 81. The ALMA staff had also used the telescope to observe two of the spectral lines of carbon monoxide. Simon was able to use the wavelengths of the spectral lines, which depend on the speed of the gas,[‡‡] to investigate the flow of gas within the galaxy.[6] He discovered that the giant clouds of gas and dust are part of a spinning disk. This is not too surprising – our own galaxy is spinning too – but the surprise was how fast it is spinning: five times faster than our galaxy. A whirling disk crammed full of fireworks – a Catherine wheel spinning in space.

One thing I discovered myself from some calculations I did based on Simon's measurements is that the disk is unstable. Within about 30 million years, it will have fallen to pieces. The birth of this galaxy is occurring in a spinning unstable disk, which appears to be already breaking up into a handful of gigantic clouds of gas and dust.

[‡‡] The Doppler shift, named after the scientist who discovered it. The change in the wavelength or frequency (pitch) of a wave is proportional to the speed of the object that is generating the wave. The best everyday example is the change in the pitch of an ambulance siren when an ambulance passes you in the street, as it switches from approaching to receding.

We are looking at the birth of a galaxy, seeing it as it happens, but what will happen next? Based on what we already know about galaxies, we can make some predictions. The stellar birth rate in SDP 81 is so high that eventually all the gas will have been turned into stars. All the short-lived stars will quickly die. Eventually, the galaxy will contain only old stars and very little gas, and the breakup of its disk means that the stars will be moving in every direction like a swarm of bees– the essential characteristics of an elliptical galaxy. It seems likely that SDP 81, whose birth 11.6 billion years ago we are witnessing thanks to the speed of light, will turn into one of those red-and-dead galaxies we see around us today.[§§] Ironically, SDP 81 will probably turn into a galaxy very like the one we are using to magnify its images.

This isn't a complete answer, of course. The birth of SDP 81 is in front of our (sub-mm) eyes. But the picture is a snapshot; we are seeing SDP 81 rather than watching it. We are catching it at just one moment during its birth. SDP 81 might be evolving quickly in cosmic terms, but it will still take much longer than all recorded human history before its disk has collapsed completely. SDP 81 is also only one galaxy and possibly the births of all galaxies are not the same.

But when we looked at the picture, it still felt pretty good.

[§§] The ages of the stellar populations in present-day elliptical galaxies are about 11 billion years.[7] These galaxies were therefore probably born about two billion years after the big bang, which agrees well with the idea that when we look at SDP 81 we are watching the birth of an elliptical galaxy.

CHAPTER 12

Andromeda Dreams

B Y THE END OF the Science Demonstration Phase, we had only received 3% of the data for the *Herschel* ATLAS. For several years, there was a piece of paper taped to the shelf above my desk with the dates scheduled for all our observations. The data for our three small fields on the celestial equator arrived at regular intervals because these fields were close to the ecliptic plane,* which *Herschel* could see for much of the year. But our big fields near the north and south Galactic poles were only visible twice a year, so we wouldn't see any data for months, and then suddenly *Herschel* would be observing our fields every day, and the data would rush in.

When it came in, the two people in Cardiff who mostly dealt with it were my postdocs Robbie Auld and Ali Dariush, a young Iranian astronomer who had joined the team after launch. Robbie and Ali got rid of the jumps and glitches, ran the program that removed the temperature drifts and fixed the occasional cooler burp. Every now and then they would transform the accumulated timeline data into a sub-mm picture.

The pictures were always less interesting than the one we had made during the Science Demonstration Phase, which had contained wisps and billowing sheets from the interstellar dust in our own galaxy (Figure 10.2). The new pictures, which were much farther from the Milky Way, always showed the same thing: tens of thousands of tiny blobs, distant sub-mm sources – pointillist pictures, with, occasionally, if we were lucky, a bright nearby galaxy to give the picture a bit of interest (Figure 12.1). While my group in Cardiff dealt with the data from SPIRE, groups in Edinburgh and at the Open University dealt with the data from the other camera, PACS. And Loretta and Steve and their group in Nottingham undertook the difficult job of finding the galaxies that were the source of the sub-mm radiation.

As the data flooded in, we started to do research with the images we had already made. When Loretta and I asked the team to propose projects that we could do with the data from the Science Demonstration Phase, we received 29 separate proposals. In choosing who would lead these projects, Loretta and I always favoured the postdocs in the data-reduction team.

---

* The ecliptic plane is the plane in which the planets orbit the Sun.

 DOI: 10.1201/9781003195290-12

FIGURE 12.1   A sub-mm picture of the sky near the north Galactic pole. The inset panel at the top right shows a close-up of a small part of the picture. The barely-visbile speckles in this panel around the sub-mm image of the nearby galaxy NGC 4725 show a small fraction of the 150,000 sources we detected in the whole picture. The left-hand inset panel shows an image of NGC 4725 in the visible waveband – more detail and much prettier.

Credit: ESA and the *Herschel* ATLAS team

For a postdoc, a big project like ours is always a mixed blessing. It may be edifying to be part of a team that is making something that is going to be used by a future generation of astronomers, but all that time spent reducing data is time that can't be used for the one thing that will get you your next postdoc contract: writing papers. We thought we owed the postdocs for all the work they were putting in – with the graft comes the reward.

Strangely, at that time I didn't have much to do myself. SPIRE worked so well that the data was very easy to reduce. And Robbie and Ali were handling this efficiently without needing any help from me – and I had been spending so much time doing other things I wasn't much use at data reduction anyway. The *Herschel*-ATLAS was the big survey that Loretta and I had dreamed up in my office a decade before, but now that it had started I really didn't have very much to do. I had time on my hands.

I was still part of Matt Griffin's team, the team that had built SPIRE. This team had some of the big chunk of observing time set aside for the instrument builders, the guaranteed time (GT). I had helped plan the GT observing programme (Chapter 6), but after we had been awarded the time for the *Herschel* ATLAS[†] I had lost interest. But every day now I

[†]   The *Herschel* ATLAS was carried out in the open time, the observing time available to everyone.

was seeing some of the results of the GT observations. The pictures were enthralling. Most of them were not like the pictures from our survey, pictures of tens of thousands of little blobs, but pictures of individual objects, the first in a new waveband. Loretta, Steve and most of the others in the *Herschel* ATLAS team were not part of Matt's GT club, but I was able to work on the GT projects as well. Some things are hard to resist.

In December 2009, I was in a bar in Madrid chatting to another astronomer. We had both been attending the first big *Herschel* conference, the conference organised by Göran Pilbratt to discuss some of the very early results. Maarten Baes, a Belgian astronomer, was a member of the PACS team, which like Matt's SPIRE team had been given a big slab of GT as a reward for building the instrument. As we chatted, Maarten and I realised there was one object missing from the observing schedule, which everyone seemed to have forgotten.

I didn't know much about the Andromeda galaxy at the time. The closest big galaxy to our own, Andromeda is one of the few it is possible to see with the naked eye. I have spent my whole career observing objects that nobody can see. When someone asks me whether some bright object is a planet or a star, I usually reply that no, I don't know what it is, the night sky is for amateurs.

But it still seemed a little strange to Maarten and me that there were no plans to observe Andromeda. For an observer, close is always good because it allows you to see more detail. It seemed ridiculous to us to spend a billion euros launching a state-of-the-art space telescope and not use it to take a picture of the biggest nearby galaxy. We also realised that Andromeda was so bright it wouldn't take much observing time. We knew that both teams still had a lot of GT in the bank, observing time for which there were no targets yet. We made a deal that once we got home we would ask our teams for some time to observe Andromeda. It would only require a tiny fraction of the GT that was left.

We calculated that it would require only eight hours of observing time from each of the teams to get some nice pictures of the galaxy with both cameras. Maarten got his eight hours from the PACS team easily enough. For my eight hours with SPIRE, I needed to ask the leaders of the group that Matt Griffin had set up to plan the GT programme for nearby galaxies (Chapter 6): Specialist Astronomy Group 2, SAG2 for short. I didn't think there would be a problem. I knew that SAG2 still had well over 100 hours of unspent GT.

Sue Madden didn't agree. One of the two co-leaders of SAG2, Sue is a scientist I respect, but we didn't agree about Andromeda. Possibly she thought my argument that we should observe it because it was the closest large galaxy was, as an observing proposal, a little on the short side. I have never talked to her about it, but my guess is that she thought, as scientists, we should never observe anything without a clear scientific purpose; we should never observe something simply because it is there. One of Sue's big research interests is dwarf galaxies. These tiny galaxies are full of dark matter, one of the big mysteries of modern astronomy. Observations of them might tell us something about dark matter. It wouldn't take much time to observe Andromeda, but that would be eight hours we couldn't use to observe dwarf galaxies. I thought… *Well, it is the closest big galaxy, Sue!*

Sue did agree, though, to hold a vote. She would ask everyone in SAG2 to vote on whether we should use some of our remaining GT to observe Andromeda. I was still not very worried.

By the time of the vote, I had begun to feel the charisma of the galaxy. Andromeda, after all, is a galaxy that even an amateur astronomer can observe – you can see it! It must have been part of every human culture since the first humanoids became aware of the night sky. Its name, which is the same as the constellation in which it is found, dates back to Greek times. Andromeda in Greek mythology was the daughter of Cepheus, the king of Aethiopia, and his wife Cassiopiea. Cassiopiea made the big mistake of boasting that she was more beautiful than the sea nymphs, as one does, and as a punishment Poseidon sent a sea monster to ravage the coast of Aethiopeia. Andromeda was chained to a rock as a sacrifice to the sea monster and was rescued in the nick of time by Perseus, who married her and took her to Greece as his queen.

It is this constellation that for some reason seems to have caught the imagination. I can think of only two books with a constellation in the title, and in both cases it is Andromeda. Both books, *A for Andromeda* and *The Andromeda Strain*, are about a threat to the Earth, in one case a radio signal that contains the information for constructing a dangerous alien creature (spoiler: don't do it!) and in the other a microorganism that threatens to wipe out all life on Earth. The galaxy itself is not the threat,‡ but its presence may be the reason for the appeal of the constellation.

I had also become more interested in the galaxy as I learned more about its place in the history of astronomy. It entered the historical record in CE 964 when the Persian astronomer Abd al-Rahman al-Sufi described it in his *Book of Fixed Stars* as a 'nebulous smear'.[1] After this not very glamorous entrance, it was included by Charles Messier in his list of the irritating objects to avoid when searching for comets (Chapter 7; it is Messier 31 - M31). And a little while later it was observed by the great man himself, William Herschel, who noticed that the centre of the galaxy is redder than its outskirts.[2]

Everyone's favourite picture of the galaxy was taken by Robert Gendler. Robert is an amateur – I am not really a snob – and Andromeda is so bright that he only needed a small telescope, which he set up on the driveway of his home in Massachusetts. Andromeda is only two million light-years away, so, unlike the galaxy in the last chapter, it almost certainly looks pretty much the same as it would today two million years later.

Robert's picture shows very nicely the difference in the colour noticed by William Herschel (Figure 12.2). The central stellar bulge is a different colour to the surrounding disk, although in Robert's picture it appears more yellow than red. His picture also shows two other galaxies: the small galaxy below Andromeda, NGC 205, which was discovered by Caroline Herschel, and the small round galaxy above it, Messier 32, which was discovered by Messier. Both are dwarf galaxies, gravitationally attached to the larger galaxy, part of Andromeda's retine of dwarfs.

---

‡   It is expected to collide with our own galaxy in about five billion years. According to some computer simulations, the two galaxies will eventually merge into a single galaxy.

FIGURE 12.2   Everyone's favourite picture of Andromeda.

Credit: ESA, Robert Gendler

In modern times, the galaxy has continued to play an important role in the history of astronomy. Early in the twentieth century, Edwin Hubble noticed there were some stars in the Andromeda nebula (as it then was) that seemed to vary in the same way as the Cepheid variable stars in our galaxy do. He was able to use their properties to estimate the distance of the nebula, showing it must be a galaxy like our own.[3] He realised that if the other nebulae he could see on his photographic plates were also galaxies like Andromeda, the fainter ones must be billions of light-years away – and with that insight, the scale of the universe increased by a factor of about one million.[§]

A little later, the German-American astronomer Walter Baade discovered the reason for the different colours. When the Second World War broke out, Baade was working at the Mount Wilson Observatory in California. As an enemy alien, he was kept under surveillance and his movements restricted. For him the silver lining, because all the American astronomers were away in the military, was unlimited observing time, and also the perfect observing conditions because of the wartime blackout. In these exceptional conditions, he was able to use the 100-inch telescope to separate the blur of Andromeda's central bulge into its individual stars, which he showed are mostly low-mass, red stars. In the bulge, there are very few of the short-lived massive blue stars that are present when the stellar birth rate is high. These in Andromeda are mostly in the disk, which explains the blue colours there.

[§]   There is an account of this discovery in my book *Origins*.

And on top of all these reasons why Andromeda had burrowed its way into my affections, I had finally realised its scientific importance. It is not just that it is the nearest big galaxy. It is the *only* big nearby one.

Andromeda and our own galaxy are both in the Local Group, a small undistinguished group of galaxies, which a cosmic tourist guide (The Lonely Universe?) would probably describe, if it bothered to mention it at all, as 'unremarkable'. The galaxies in the Local Group are important ones for astronomers because they are the only ones close enough for us to see the individual stars, which makes possible many more investigatory techniques than for galaxies further away. There are almost 100 galaxies in the Local Group, but out of these 97 are dwarfs. There are only three spiral galaxies in the Local Group and one of these has a mass much smaller than our own galaxy.⁵ Of the remaining two, one is our own. The other is Andromeda.

There are a couple of reasons why a big nearby spiral galaxy is important. The first is that it is always important to find a second example of something, otherwise you might get a misleading impression. Suppose an extraterrestrial species visited the Earth and, for some reason, there was only one human left. They might get a very peculiar view of the human race. A second example makes it possible to begin to distinguish which are common characteristics and which are individual variations. As it happens, we already know that Andromeda is different from our own galaxy in some interesting ways. It has a much larger central bulge but less prominent spiral arms, although it's hard to be sure about the arms because its highly inclined disk makes them hard to see.[4]

The other reason is that it's often easier to find out things about spiral galaxies by studying a nearby one than by studying our own. The problem with studying our own galaxy is our location. We can only study it from the inside, which means it has been surprisingly difficult to find out even some very basic things about it, such as the number of its spiral arms.

So by the day of the vote, I wanted to win. I was not worried. Andromeda was an iconic galaxy, it was routinely observed by every new observatory, and it would only require a small amount of observing time. There might be a few eccentrics like Sue who wanted to reserve the observing time for something else, but I couldn't believe many people would vote against observing the nearest big galaxy.

When I got in that morning, Walter Gear, who was the co-leader with Sue of SAG2, came round to see me. Walter, who agreed with me about Andromeda, said that every French astronomer in SAG2 seemed to have voted against observing it. It was now only three hours until the deadline. He said that he thought we would lose unless I did something. After he left, I sent an email to everyone I thought might not have voted yet, telling them there was a chance that *Herschel*, a billion-euro telescope, might end up not observing the nearest big galaxy. When the votes were counted at midday, we won by two votes.

We had the observing time. But the next visibility window for Andromeda would not be for months. We had to wait.

---

⁵   Messier 33. It is sometimes called the Triangulum galaxy because it is in that constellation.

And so it was back to the survey. Most of my time in 2010 was not spent thinking about Andromeda or gravitational lenses or looking at the latest exciting *Herschel* picture. It was spent in preparation for our first big data release. Data releases are never fun, but they are a fact of life in big astronomy projects.

I had promised ESA in our proposal that we would release our finished images and source catalogues to the rest of the community at regular intervals. Until the data is released, it is effectively the private property of the team that made the observations, who can pick through it for all the most interesting scientific results. The data release is the moment you must start sharing, and for that reason some teams have become notorious for delaying it. We didn't want to be known as one of those teams, and anyway, we had been given the opportunity to carry out the first big survey in the sub-mm waveband, and we really did want to give other astronomers the excitement of looking at the images and playing with the catalogues of sources.

Loretta and I had decided, as our first data release, to release the images and catalogues for the three small fields because the observations for these would be finished first. There was a glitch in our plans when the *Herschel* star-tracker, which kept the telescope pointing in the right direction, failed. The failure ruined the observations of half of one of the fields, but after Loretta and I discussed it, we decided to go ahead anyway and release the images and catalogues for the rest of the fields. A data release is always a monumental amount of work, slow, painstaking and a bit boring. It is not just a matter of measuring the position and radiation strength for hundreds of thousands of sources. Any measurement, for it to be scientifically useful, also needs an error, the uncertainty in the measurement, and this is often as difficult to estimate as making the measurement itself. We also planned to release catalogues of the galaxies we had identified as the source of the sub-mm radiation, which turned out to be one of the most challenging things we had to do.

A data release is always harder when you are working with a new type of instrument. Ironically, one of our worst problems was that the instrument built by Matt Griffin and his team was too good. SPIRE was so sensitive there were sources everywhere. There were so many sources in the images (Figure 12.1) that they often merged, making it hard to tell where one stopped and another started. We spent years trying to solve this problem.

It is probably always the case that most of the work is done by a few people. There were about 200 people in the *Herschel*-ATLAS team, but there were only about 10 in the data-reduction group, which was responsible for preparing the data release. The membership changed over the years. At this time, the data-reduction for SPIRE was done by my group in Cardiff. The data reduction for PACS was mostly done by Eduardo Ibar, a postdoc in Edinburgh, and Ros Hopwood, a PhD student at the Open University. The group in Nottingham – Loretta, Steve, their postdoc, Dan Smith, and their PhD student, Emma Rigby – did everything else.

The Nottingham group had some of the hardest jobs. Steve had written the computer program to find the sources in the images. When we had finished making an image in

Cardiff, we would send it over to Nottingham, and Emma would run Steve's program, which would find the sources in the image, measuring the position and radiation strength of each source. The Nottingham group also had the job of identifying the galaxies that were the source of the sub-mm radiation. Loretta and Steve were regular university academics with a full teaching load, so much of their work on the data release had to be done in the evenings. They would spend their days lecturing and dealing with undergraduates, and then come home and work on the survey. After dinner, they would look at what Emma and Dan had done during the day, do all the checks on the latest version of the catalogues, and think about what needed to be done next. Steve would also need to fix the bugs we kept finding in his source-finding program. It is not unusual in astronomy to work in the evenings and at weekends, but by late 2010 Loretta and Steve had been doing it for over a year.

Our data releases sometimes felt like they had been designed by some sadistic psychologist to stress test a team. The worst part was the nearby galaxies. Most of the sources in our images looked pretty much the same, with every blob very similar to every other blob. All we needed to do was run Steve's program, and it would spit out the position and strength of the radiation of each source. But in each image there were always a few nearby galaxies, ones that were so close that we didn't just see an anonymous blob but instead something with detail like the sub-mm picture of NGC 4725 in Figure 12.1. The problem was that, for one of these nearby galaxies, we couldn't just rely on Steve's program to measure the strength of the sub-mm radiation from it. Instead, we needed to add up all the sub-mm radiation in the image that we thought might be coming from the galaxy, which in practice meant defining an ellipse that enclosed as much as possible of the galaxy's radiation but as little as possible of the radiation from the other sources (Figure 12.3). It was annoyingly subjective. In the final stages of every data release, in which we might be producing a catalogue of over 100,000 sources, we would often spend days staring at a computer screen, arguing about the sizes and shapes of the ellipses to draw around a handful of nearby galaxies.

By the end of 2010, we were getting close to the first data release. But then the stress really began to mount. Doing a data release is like jumping off a cliff – once you've done it, there's no going back. If there is anything wrong in one of the data products, it must have been a mistake by someone in the data-reduction team. And out there in the community, if someone writes a paper that turns out to be wrong because it is based on this data product, it is not their fault – it's *yours*. There is also the stress from working to a deadline. I am not sure we ever met any of the deadlines we set ourselves, but missing deadlines is stressful as well.

In December 2010, I was able to get away from this for a while. Andromeda had finally appeared in the *Herschel* observing schedule. We were lucky. ESA had been contacted by

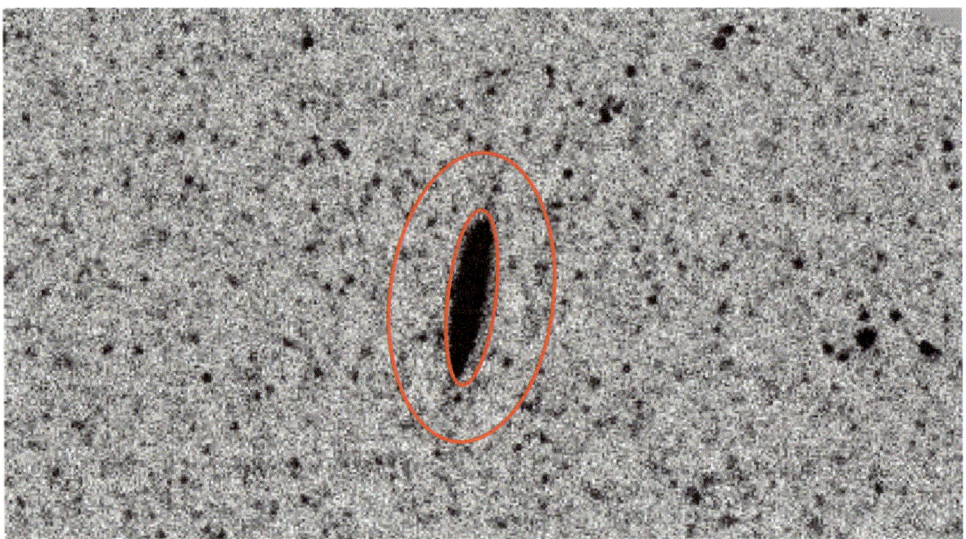

FIGURE 12.3   The nearby galaxy problem. For the sub-mm image of a nearby galaxy, like the one in the picture, we needed to measure the sub-mm radiation from the galaxy while not including any from any of the other sub-mm sources, the blobs. In practice, this meant looking at the image on a computer screen and defining an ellipse that enclosed as much as possible of the galaxy's radiation but as few as possible of the other sources. The outer red ellipse is clearly a bad one because it encloses some of the blobs. The inner one seems about right. We spent days arguing about these ellipses.

Credit: ESA and the *Herschel* ATLAS team

the producers of the BBC programme *Stargazing Live*. They wanted to show a *Herschel* picture of Andromeda in one of their episodes, so ESA had brought forward our observations.

I had already decided that the observations would be given to the new PhD student we had taken on just before *Herschel* was launched. From south London, Matt Smith was the first member of his family to go to university. His father, an engineer with the telecoms company BT, was an amateur astronomer and sometimes set up his telescope on top of the BT Tower in central London. Matt remembers watching a solar eclipse there. Before he came to us, Matt had done a physics degree in Cambridge, and since he had come to Cardiff he had been doing a lot of unselfish work helping the postdocs in reducing the data for our big survey. I thought the Andromeda observations could be the individual project he needed for his PhD thesis.

In agreeing the schedule, ESA and the BBC had not bothered to think about the personal lives of astronomers. When we received the data from the spacecraft, Matt and I were both home for the Christmas holidays. Then we learned that the episode of *Stargazing Live* would be broadcast on 3 January. Matt spent most of his Christmas holiday holed up in his teenage bedroom, logged onto the computers back in Cardiff as he reduced the Andromeda data. By the end of the holidays, he had maxed out his parents' internet account, but in front of him on the screen was the first ever sub-mm image of the nearest big galaxy. A few days later it was shown on the BBC.

When I saw it for the first time (Figure 12.4), surrounded by the detritus of Christmas, I saw it was nothing like Robert's picture, which I have shown again below (Figure 12.5). Robert's picture, in the visible waveband, shows the starlight. Our sub-mm picture shows the radiation from the dust grains. The central stellar bulge, which is so obvious in Robert's picture, in ours has completely disappeared. The sub-mm picture is dominated by the swirls, spurs, loops and rings in the disk, especially the bright ring out towards the edge of the galaxy.

The beauty of pictures in the visible waveband is the result of the short-lived massive stars – why the *Hubble* pictures of spirals are so striking but the pictures of ellipticals, which contain very few of these stars, are a little dull. These massive stars have short lives but merry ones: a star with a mass ten times greater than the Sun burns through its nuclear fuel, in a prodigal outburst of radiation, 1,000 times faster. Most of this radiation, though,

FIGURE 12.4   Our sub-mm picture of Andromeda.

Credit: ESA and HELGA team

FIGURE 12.5   The picture in the visible waveband again.

Credit: ESA, Robert Gendler and the HELGA team

is hidden from view. The birthplace of a star is a cloud of gas and dust. A high-mass star has such a short life that it spends most of it, sometimes all of it, cocooned in its birth cloud, where the dust often hides it completely from the view of optical telescopes. The energy in the starlight, though, is impossible to hide completely. The absorbed starlight heats up the dust grains, which emit sub-mm radiation, transforming the energy in the light into the same amount in sub-mm radiation.

It is this hidden starlight that explains the dramatic differences between the pictures. The stellar bulge is very prominent in Robert's picture because it contains a lot of stars, but these are all low-mass long-lived stars. Today in Andromeda, there is very little gas and dust in the centre, and the stellar birth rate there is very low. The disk is where stars are still being born, and the large number of short-lived, massive blue stars there explains its colour. Most of these massive stars, however, are completely hidden by the dust, especially those in the ring of gas and dust at the edge of the disk. The ring is invisible in the optical picture because most of the light from the massive stars in the ring is hidden by the dust.** We can see the ring in our picture because the starlight heats the dust, which emits sub-mm radiation.

Robert's picture is beautiful, but it is a little like a profile on social media. It does not tell the whole truth.

This ring is one of the biggest mysteries about Andromeda. Why is there one there and not in our own galaxy? For a while, astronomers thought they might know the answer. It is thought that the orbit of M32, the dwarf visible just above the stellar bulge (Figure 12.5), may take it through the disk of the big galaxy. The South African astronomer David Block and his colleagues suggested this might be the explanation of the ring.[5] They suggested that the passage of the dwarf through the disk had generated a wave that travelled out through the gas in the disk, like the circular waves generated when a stone is dropped into water. They suggested that the motion of the wave out through the disk might – we are only just beginning to understand how stars are born (Chapter 9) – have caused the gas clouds in the disk to collapse and form stars. If all this is true, Andromeda's ring is the crest of that wave, a tsunami moving through the disk seen frozen in our *Herschel* snapshot.

Unfortunately, because it is such a neat story, there is now evidence it is wrong. If the story is correct, the wave's journey out through the disk should have produced a sequence of star birth. The stars further out in the disk should therefore be younger, on average, than the stars closer to the centre. And this is where our ability to see the individual stars in Andromeda becomes important. Around the time we observed Andromeda with *Herschel*, a big international team led by the American astronomer Julianne Dalcanton used *Hubble* to observe 100 million stars in the galaxy. Estimating the ages of the stars from their colours, the team found that the ages of the stars are pretty much the same everywhere in the disk.[6] Andromeda's ring remains a mystery. Our picture of it is beautiful, though.

By this stage, Sue herself had become enthusiastic about the project. She had been right about one thing. We didn't have a clear scientific goal in making the observations. Even the ring had been seen before. When Matt and I looked at an old picture of Andromeda taken with the Infrared Astronomy Satellite in the 80s, there it was – a little blurry and not with the exquisite, filigreed structure seen in our picture, but it was there.

When I hunted around to find things to do with the data, all I could think to do was to measure the temperature of the dust. We had taken pictures of Andromeda at five different wavelengths (Figure 12.6), and when Matt and I compared them we thought we could see variations in the dust temperature. The ring seemed to stand out best at 70 micrometres, the shortest wavelength. This made sense because the warmer the dust, the shorter the wavelength at which its radiation is strongest (Chapter 2), and it seemed likely to us that the dust would be warmest in the ring because of the radiation from the massive stars in the ring.

I suggested to Matt that he use the five images to investigate how the temperature of the dust varies through the galaxy. There is a precise mathematical relationship between the strength of the radiation from the dust, the wavelength and the temperature of the

---

** The dust can be seen in Robert's picture as the faint dark bands.

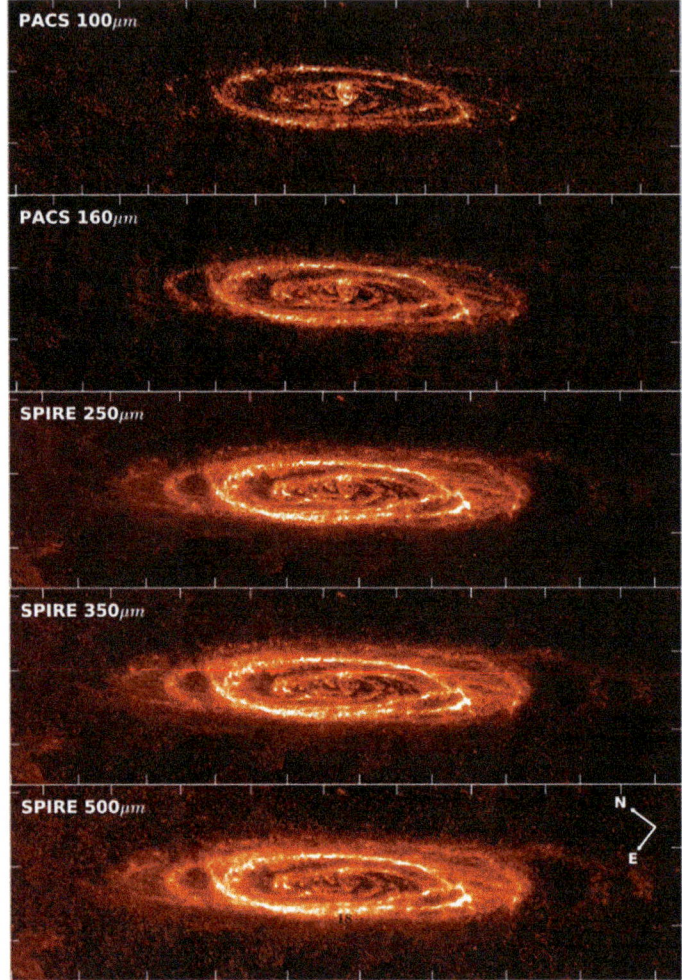

FIGURE 12.6   Our five images of Andromeda, at wavelengths from top to bottom of 70, 160, 250, 350 and 500 micrometres. The image shown in Figure 12.4 is the one at 250 micrometres, our most sensitive image.

Credit: Matt Smith

dust. Given measurements of the radiation strength at five different wavelengths, it's fairly straightforward to adjust the temperature of the dust until the relationship matches the measurements – and so estimate the temperature of the dust. Back in the 80s, I had done this for galaxies as a whole (Chapter 2). I reckoned that Matt would now be able to do the same thing at several thousand separate points in Andromeda, allowing him to investigate, in great detail, how the temperature of the dust changes through the galaxy.

As we discussed it, we realised there was one other thing he could do. On either side of the peak, the strength of the radiation falls off. The steepeness of the decline on the long-wavelength side, the 'slope', depends on the chemical composition of the dust grains and on their sizes and their shapes, although nobody is sure – we still know so little about dust – which of these has the strongest effect. Neither of us expected the dust grains to be different in one part of Andromeda from another, so we didn't expect to see any variation in the slope, but we thought Matt might as well measure it.

A few weeks later, Matt came back to me. The first thing he showed me was his map of the temperature of the dust.[7] There on the screen was the most detailed map ever made of the temperature of the dust in a galaxy (Figure 12.7). Nevertheless, it didn't show anything very surprising. His map showed that the temperature of the dust in Andromeda is mostly between 16 and 18 kelvins, with the temperature being a little higher in the ring, which we had already guessed. The only thing that was surprising was the temperature in the centre of the galaxy, which was about 30 kelvins. There isn't a lot of dust in the central region of Andromeda, but the dust grains there are warmer than in the rest of the galaxy. Matt had already thought of the explanation. Dust grains are heated by starlight and the stars in the bulge are closer together than in the disk, so the intensity of the starlight there is higher – and so is the temperature of the dust.

Then Matt showed me his map of the slope. We hadn't expected this to vary, but there on the screen it did, falling gradually from very high values (a steep slope) in the centre to much lower values in the outskirts of the galaxy (Figure 12.8). We realised that we had been wrong in our assumption. The dust is not the same everywhere in Andromeda. There must

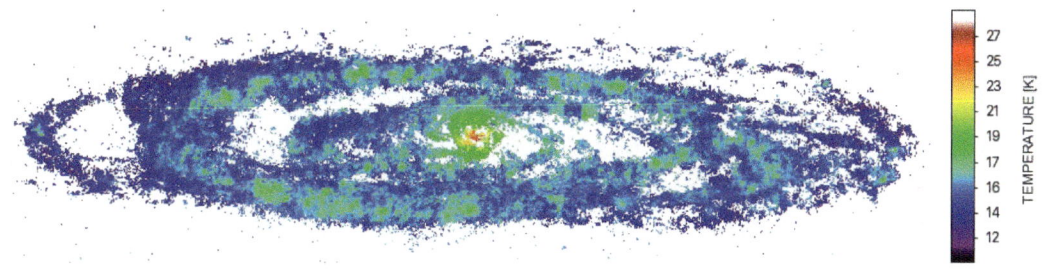

FIGURE 12.7   Matt's map of the temperature of the dust in Andromeda. The bar shows the colour code.

Credit: Matt Smith

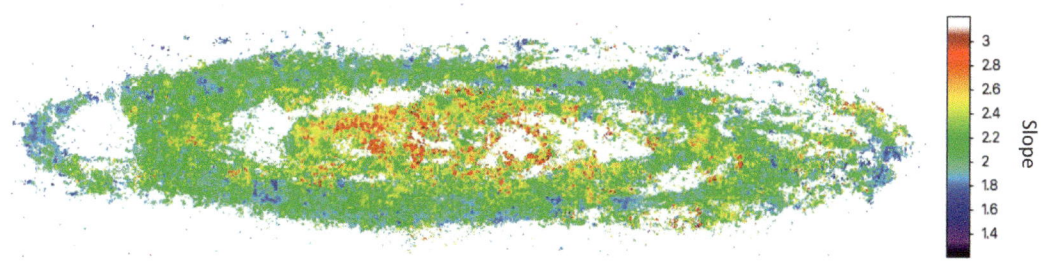

FIGURE 12.8   Matt's map of the slope. The bar shows the colour code.

Credit: Matt Smith

be some systematic change in the properties of the dust from the centre of the galaxy out to its edge. A decade later, we still don't understand enough about the physics and chemistry of dust to know what this change is. Are the dust grains in the centre smaller than those farther out? Are the grains in the centre mostly made of carbon compounds and those further out of silicon compounds? Or is it something to do with the shapes of the dust grains? We just don't know. In the last decade, similar radial variations in the properties of dust have been seen in other galaxies,[8] but we were the first to see it.

A discovery like this is a wonderful feeling. Make a bigger discovery, and almost certainly the moment you make it you are thinking about what it will mean for your career. This was one of the purer ones. It wasn't going to benefit anyone's career; there were probably only 100 people in the world who would even care about it. But Matt and I were the first two people, out of the nine billion people on the planet, to know that the properties of interstellar dust often change from the centre of a galaxy to its outskirts. And it was an unexpected discovery. We hadn't set out to make it. We had just stumbled across it – something new about the universe.

We went on to use the Andromeda images for other projects. We used the *Herschel* images to estimate the rate at which stars are being born in Andromeda. The stellar birth rate there is several times less than in our own galaxy.[9] In that, too, Andromeda is not quite our twin. We also used the images to compile a catalogue of clouds, which we then used to trace out the galaxy's faint very-hard-to-see spiral structure.[10] And the beautiful *Herschel* images of Andromeda continue to be used for other people's projects.

But this was the discovery I remember, the one that came out of the blue.

CHAPTER 13

Woomph!

I N EARLY 2011 WE stepped back from the edge. We decided to delay releasing our catalogues and images. We kept them to ourselves for a bit longer.

We did have a reason. The camera built by Matt Griffin and his team was so sensitive that there was a source virtually everywhere we looked – sources on sources. The program that Steve Maddox had written to find all the sources did a fairly good job at separating the overlapping ones, but it wasn't perfect, and in early 2011 he thought of a better way of doing it. He claimed his new method would not only separate the sources better but would find many more of them, which would be a very good thing. But it would require months of extra work and would delay the data release, which would not be ideal. On balance, Loretta and I thought it was just about worth it, and we postponed the data release.

And it did give us more time to play with the data ourselves.

When I look back, I see that our team published over 30 scientific papers based on those early observations during the Science Demonstration Phase, including, despite Göran's crazy deadline, eight in the special *Herschel* issue of *Astronomy and Astrophysics*. One of these was Simon's paper describing our discovery of recent cosmic evolution.[1] While everyone knows that galaxies ten billion years ago were very different from those today, we had discovered that even one billion years ago galaxies were different, which in cosmic terms is almost yesterday.

About this time, we realised that our survey – in an unexplored waveband – gave us the chance to perform the first fair test of the computer simulations that are now one of the main methods used by astronomers to understand the universe.

The simulations, which are generated by computer programs, are impressive. Tens-of-thousands of lines of code written by large teams of astronomers, these computer programs produce simulations of the history of the universe, starting a fraction of a second after the big bang and ending up 14 billion years later with a model of the universe we see around us today – all compressed into a few months of computer time. They generate movie sequences of the evolution of galaxies, showing them as they grow, metamorphose and merge with other galaxies, which are often of such compelling beauty it is difficult to avoid the conclusion the simulators must have got something right. They also end up with galaxies in the universe today that do look like real galaxies (Figure 13.1).

DOI: 10.1201/9781003195290-13

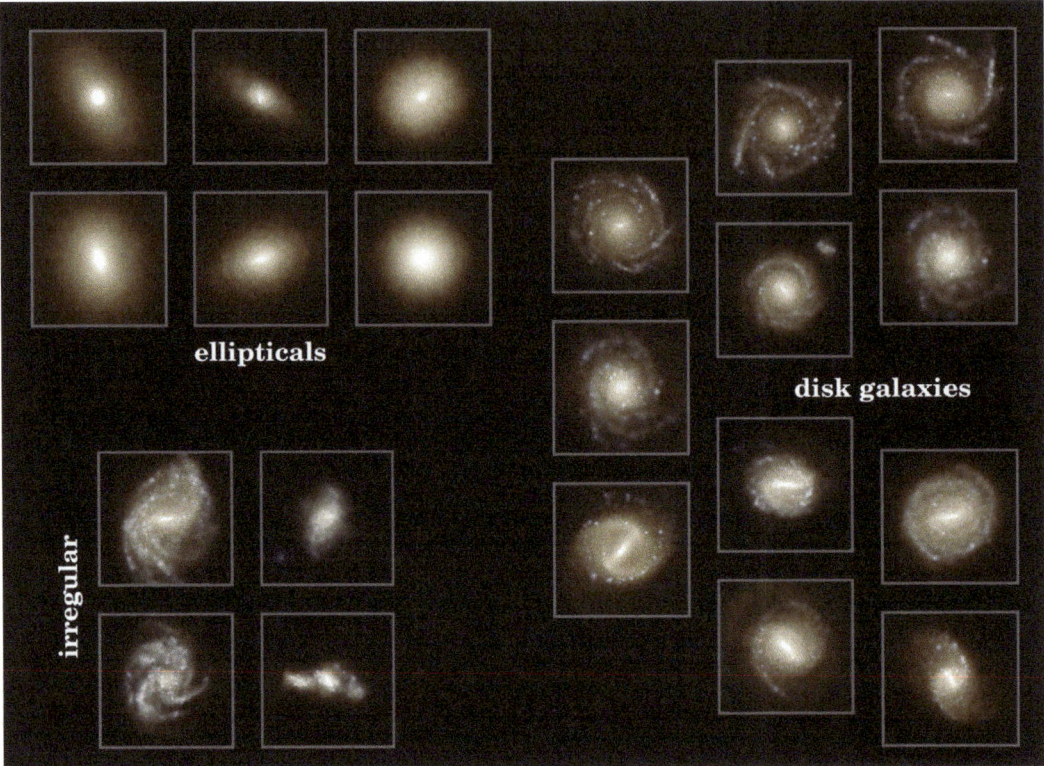

FIGURE 13.1 Pictures of some artificial galaxies generated in a computer simulation of the universe.[2]

Credit: Illustris Collaboration/Illustris Simulation

Look under the bonnet, however, and the models don't look quite as impressive. Although there are some shiny parts in the engine, they often look as if they are held together by bits of string and chewing gum – something that, to be fair, the simulators usually acknowledge. The big challenge for the simulators is that there are still big gaps in our understanding of how galaxies evolve.

The way the simulators fix this is by inserting a few lines of code in the program that make something happen even if we don't understand why it happens. A good analogy is to imagine trying to construct a computer simulation of the flight of a plane even if you don't understand how such a large object can stay in the air. This is really the perfect analogy because it's easy enough to create a simulation of plane flight without having any understanding of the physics of how it happens in the real world. All you need is a line in your program that instructs the computer to make the plane in the simulation take off when it reaches a certain speed on the runway.

One of the biggest gaps in our understanding of galaxies is the explanation of one of the most obvious characteristics of the galaxy population: the green valley.

As I described in Chapter 11, the galaxies found in a survey in the visible waveband mostly fall in two distinct regions of a diagram of optical luminosity versus colour (Figure 11.1). The galaxies that contain a large amount of gas and have a high stellar birthrate mostly fall in a region of the diagram that astronomers have named the 'blue cloud'. The galaxies without much gas and with a low stellar birthrate fall in a strip that astronomers call the 'red sequence'. The area in the diagram between these two regions, where there are very few galaxies, is the 'green valley'.

The green valley is a problem. As a galaxy runs out of gas, its stellar birthrate will gradually fall. As the stellar birthrate falls, the number of blue short-lived massive stars also falls. As a result, the colour of the galaxy also changes and it moves from the blue cloud, across the green valley, to the red sequence. The big problem is that the natural pace at which this happens is quite slow, and if this was all that happened in the evolution of a galaxy, we should see many more galaxies in the green valley, as they slowly move from the blue cloud to the red sequence, than we do.

Astronomers have spent a lot of time trying to think of processes that might reduce the gas content of a galaxy fast enough for it to move rapidly across the green valley – fast enough that we see few galaxies there. There are many ideas. The most popular one today, although I can think of at least four others, is that it has something to do with the super-massive black holes at the centres of galaxies (Chapter 6). Supply one of these sleeping monsters with gas and it will come to life, turning into an active galactic nucleus, which might be a quasar, emitting so much radiation that it is impossible to see the surrounding galaxy, or a radio galaxy squirting jets of relativistic particles into intergalactic space (Chapter 2). It seems plausible that the radiation from a quasar, for example, could expel the gas from a galaxy, leading to a rapid reduction in its stellar birthrate and hurrying it across the green valley. The problem with this idea is that the details of how this might happen – the physics – are still very unclear.

The simulators don't worry about this. They simply assume this is what happens. They insert a few lines of code in their programs that, at the required moment, reduce the stellar birthrate in the galaxy*. With these instructions, the simulation *does* produce galaxies that move across the green valley fast enough for there to be few galaxies there.

But this also means that it is not a fair test of a simulation to see whether it produces the green valley: the program has been designed so that is what it does. In the same way, it would not be a fair test of a flight-simulation program to check whether the plane takes off into the sky, since the modellers inserted a line in the program to make sure this happens. The same is true of most of what we know about galaxies. The simulators have tweaked and adjusted their programs so that the simulations generate an artificial universe as close as possible to the real universe we see around us – artificial images of galaxies that look like real galaxies, for example.

At some point it dawned on us that we were in the perfect position to test these simulations. We were carrying out a survey in a waveband that had never been explored before, and so anything we discovered couldn't have been known by the simulators when they wrote their programs. We had the chance to do the first completely fair test of the simulations. We decided to do a test of the simulations by seeing whether they predicted the rapid recent cosmic evolution discovered by Simon.

---

* I've simplified things slightly, but if any simulators are reading this, I think this is a fair statement.

When we checked, we found that the simulations did not predict that galaxies one billion years in the past were any different from those today. In the simulations, but not in the real universe, galaxies haven't changed very much during the last billion years. This made me, as an observer, quite happy.[3]

Around this time, we also began to discuss within the team the cause of Simon's result. Why were galaxies one billion years ago emitting more sub-mm radiation than they do today? The sub-mm radiation is from the dust in the galaxies, which in most cases is heated – as in Andromeda – by short-lived massive stars. We guessed that the explanation was that one billion years ago galaxies contained more gas than they do today, which led to a higher stellar birthrate and so to a larger number of short-lived massive stars. Since gas and dust are always found together, we realised that the way to test our hypothesis was to check whether galaxies one billion years ago contained more dust than they do today. When we did this*, we found it was as we had guessed. One billion years ago galaxies were emitting more sub-mm radiation because they contained more dust (and gas) than galaxies today. This did raise the question of *why* they contained more gas and dust, but there are always more questions. We were just happy to find the answer to one of them.[4]

And here's something else for the list of why dust is important. One of the most annoying things about doing research on the evolution of galaxies is that we have no direct way of measuring how much molecular hydrogen there is in a galaxy, which is the galaxy's reservoir of raw material out of which its stars are made. It is possible to estimate the depth of the reservoir from observations of carbon monoxide, but as I explained in Chapter 9 there are a lot of problems with this method. A much better way is to estimate the mass of the molecular hydrogen from sub-mm observations of the dust.[5] With our *Herschel* survey, we now had the chance to estimate the depth of this reservoir for hundreds of thousands of galaxies.

But at this time, in the middle of 2011, nobody knew where all the dust had come from.

The night sky changes infinitesimally slowly. Human civilisations rise and fall and the sky changes hardly at all. A star like the Sun has a lifetime of about ten billion years, longer than the age of the Earth; even the stars with the shortest lives last for about ten million years, much longer than humans have been walking on the planet; and it would take a billion years or so before the appearance of a galaxy changed very much.

The collapse of a massive star, though, happens in a heartbeat.

In the instant before the star's collapse, the structure of its core is like some exquisite French pastry with every layer made of a different ingredient, or in the case of a star's core a different chemical element (Figure 13.2). The outermost layer is made of the lightest element, hydrogen, the most common element in the universe. Below this is a layer of helium, the second most common although there is hardly any on Earth,** and below this are layers

---

* We were able to estimate the mass of dust in a galaxy because we had images at five wavelengths, which allowed us to estimate the temperature of the dust. It is important to know the temperature of the dust because warm dust emits more radiation than the same mass of cold dust. But if a correction is made for the effect of temperature, it is possible to estimate the mass of the dust from the strength of the sub-mm radiation.

** Its existence was only discovered from unknown spectral lines in the Sun's spectrum.

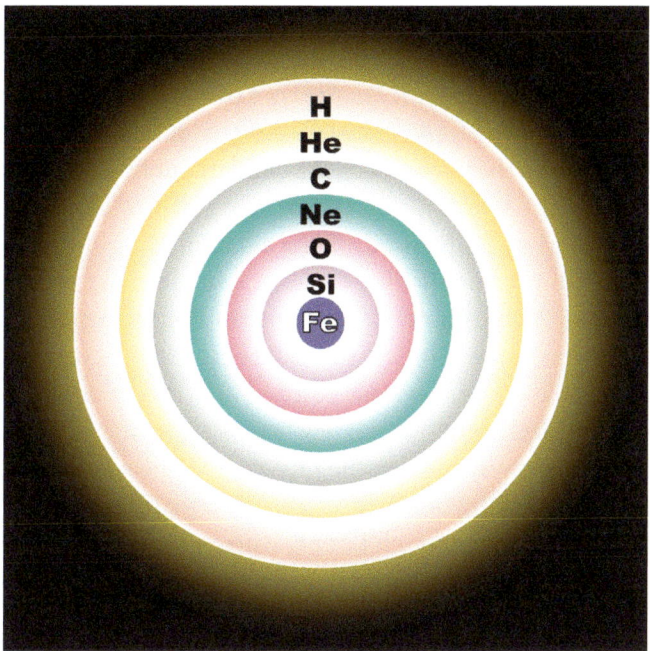

FIGURE 13.2   The layers of the core of a massive star shortly before a supernova explosion.

of carbon (pencils, diamonds, biological molecules), neon (fluorescent light bulbs), oxygen (respiration) and silicon (computer chips). And at the heart of the pastry is the most useful of all elements: iron (bridges and cars).

The iron is not much like the iron in a bridge – its density is so large that a teaspoon would weigh about the same as a large truck – but its function in the star is pretty much the same. It is the pressure in the iron that is holding up the core, stopping it from collapsing under its own weight. The pressure in the iron, though, is not the kind of pressure that stops the core of the Sun, a much lower mass star, from collapsing. The pressure that supports the Sun's core is the everyday kind, the pressure that keeps balloons and tyres inflated – the pressure exerted by fast-moving particles bouncing around. On Earth these particles are gas molecules, in the Sun's core atomic nuclei and electrons.

The ultimate energy source of all stars is nuclear fusion. In the time I take to write this sentence, about one billion tonnes of hydrogen in the Sun's core has been transformed into helium. Each time four hydrogen nuclei are fused into a helium nucleus, a tiny amount of mass is lost, mass that has been transformed into energy.* One billion tonnes sounds like a lot, but the Sun contains a lot of hydrogen. It will be about five billion years before it all

---

* An atom consists of the nucleus, made of protons and neutrons, surrounded by electrons.. Different elements have atomic nuclei with different numbers of protons and neutrons. The hydrogen nucleus consists of a single proton. The helium nucleus contains two protons and two neutrons. When four hydrogen nuclei, protons, are fused into a helium nucleus, two of the protons are turned into neutrons, with the mass of the helium nucleus being slightly less than the combined mass of the four protons. The mass that makes up the difference has been transformed into energy, the amount given by Einstein's famous formula: $E = mc^2$. In this equation, c is the speed of light, which is a very large number, so a small amount of mass has been converted into a very large amount of energy.

runs out. In the core of the high-mass star shown in the figure, nuclear fusion has now stopped completely in the iron, although it's still happening in the surrounding layers.

When all the hydrogen that is initially in the core of a star has been converted to helium, energy production temporarily stops. The star, though, is still shining. Energy is leaking out of the core, which means the particles in the core gradually lose energy and slow down. The pressure drops. The core is still just as heavy, so it starts to contract.

The only difference in what happens now from a boulder tumbling down a mountain, or my coffee cup falling onto the floor, is that the core is descending through its own gravitational field rather than through the gravitational field of something else. But the transformation of energy that happens is the same. The initial energy of the core from its position high in its own gravitational field – like a boulder perched high on a mountain – is now gradually released as the core contracts, being ultimately transformed into heat. One of the weirder consequences of the laws of physics is that the rate at which heat is generated is exactly twice the rate at which energy is leaking out of the core. The core therefore gets hotter.

When the core reaches a high enough temperature, a new type of nuclear fusion starts. The helium nuclei start combining to form carbon and oxygen nuclei.* Energy is now being produced again, which stabilises the pressure, and the core stops contracting. In a low-mass star like the Sun, the hydrogen-fusing phase will last about ten billion years. Its helium-fusing phase will last about one billion years.

Eventually, though, all the helium will be gone. If the star has a low mass (like the Sun), there will be no new energy source to replace the energy from helium fusion. Once all the hydrogen in its core has been exhausted, its outer layers will expand as it becomes a red giant star. When all the helium is gone, it will swell even more to become a supergiant star. Eventually, its outer layers will stream away, leaving its core unveiled as a white dwarf star, surrounded by one of the most beautiful objects in the universe: a planetary nebula (Chapter 8).

If a star has a high mass, like the one shown in Figure 13.2, with a mass at least eight times the mass of the Sun, this is not the end of the road. When all the helium is gone, the core, now made of carbon and oxygen, will start to contract again. The temperature of the core will increase, and when it is high enough the carbon nuclei will start to fuse. Energy will be produced again, the pressure will stabilise and the core will stop contracting.

In a high-mass star like this, the same sequence happens again and again. The core runs out of fuel; it starts to contract; it gets hotter; and eventually a new cycle of nuclear fusion starts. Each cycle takes less time. For a star with a mass 25 times that of the Sun, it will take 1,000 days to use up all the carbon but, in the final stage of nuclear fusion, it will take only five days to transform all the silicon into iron.

Once all the silicon has gone, even for a massive star this is the end of the road. Iron nuclei are the most stable nuclei, so it costs energy to fuse them rather than making any. By this stage in its life, which is shown in the figure, the star's iron heart is surrounded

---

* Two nuclei will only fuse if they get close enough, which usually doesn't happen because the electrical charge on a nucleus repels other nuclei. They will only get close enough to fuse if the two nuclei are travelling at a high enough speed towards each other to overcome the repulsive force. The higher the temperature of the gas, the higher the speeds of the particles. The electrical charge is larger on a helium nucleus than a hydrogen nucleus, so the core of a star needs to be at a higher temperature for helium fusion than hydrogen fusion.

by concentric shells, the relics of all the cycles of nuclear fusion. There are no longer any nuclear fusion reactions within the iron, but there are still fusion reactions in the surrounding shells.The shell immediately surrounding the iron is made of silicon nuclei, which are steadily fusing into iron nuclei.

The star now has a short reprieve.

One of the strange properties of electrons – neither particles nor waves, impossible to visualise, indescribable literally with any of the words from our everyday world – is their resistance if too many are confined in a small space.* In a star's core the pressure from this resistance is usually much lower than the everyday kind of pressure, which today is stopping the Sun collapse under its own weight. But after all these cycles of contraction, the mixture of iron nuclei and electrons at the centre of the high-mass star has been squeezed into a much smaller volume than the core of the Sun. As energy continues to leaks out of the iron heart of the star, the everyday kind of pressure is no longer enough to avert a collapse. But the collapse is stopped by this different kind of pressure.

The maximum mass that can be supported by this pressure is 1.4 times the mass of the Sun. The mass of iron is safely below this limit.

But in the shell surrounding the iron, more iron is being made from the fusion of silicon nuclei. The mass of iron at the centre of the star is steadily increasing.

Woomph!

The iron heart of the star collapses in less than a heartbeat. It plummets inwards at a speed that reaches 70,000 kilometres per second, 23% of the speed of light. The collapse takes only two-thousandths of a second.

As the iron collapses under its own weight, like a mountain instantaneously crumbling to the ground, energy is released, turned into heat and then into photons of radiation. The photons zap the iron nuclei, splitting them into protons and neutrons. The protons and the electrons combine into more neutrons, with the release of a flood of neutrinos, another type of particle with a mass one million times less than the mass of the electron. As the heart of the star collapses, its density increases rapidly….

Eventually, when the diameter of the collapsing material is about 30 kilometres, the pressure from the neutrons, which like electrons also resist being confined in too small a space**, stops any further collapse. The material, now mostly neutrons, has become a neutron star, and we are now, strangely, back on the human scale because this is about the distance I could cycle in a morning. The material that is now in the neutron star, though, was only a small part of the original star. With the collapse of the iron, the star's outer layers are resting on nothing. They collapse, crash down on top of the neutron star, bounce off and are thrown into space in one of the universe's most energetic events: a supernova explosion.

At least, back in the early 80s, that was what everyone thought would happen. By then, astronomers thought they understood the nuclear processes in stars well enough to predict

---

* As I explained earlier, electrons are neither particles nor waves – nor anything else from the wordage of our everyday world – but it is sometimes useful to treat them as one or the other. One of the consequences of their wave-like behaviour is that if they are confined within a volume, such as the core of a star, they exert an outwards pressure. The technical name for this is electron degeneracy pressure.

** This pressure is called neutron degeneracy pressure.

what would happen when one runs out of fuel. But there was one embarrassing problem. As I mentioned above, the computer simulations of the universe are not perfect, but they do at least produce objects that look like real galaxies. The computer simulations of a supernova explosion failed this basic test. A star collapsed readily enough in one of the simulations, but its outer layers did not then bounce off the neutron star and fly off into space – a splat rather than an explosion. Given the failure of the simulations, it would have been helpful back then to have some real supernovae to study.

But here was another problem. The last supernova in the Galaxy was discovered by the astronomer Johannes Kepler in 1604, which was five years before the invention of the telescope. So, in four centuries of bigger and better telescopes and more and more ingenious new instruments, nobody had ever been able to study a nearby supernova.

But then somebody noticed something in the sky.

In 1979, Ian Shelton had just completed a degree at the University of Manitoba and wasn't sure what he wanted to do as a career. He was interested in astronomy, but he wasn't sure he could face the years of work needed to become a professional astronomer. He was also worried that turning astronomy into a career might make him lose the sense of wonder he always felt when he looked at the night sky. As a stopgap, he decided to apply for the job of resident astronomer at the University of Toronto Southern Observatory (UTSO).

The UTSO had only a single employee, the resident astronomer, and only a single, small telescope. Part of the Las Campanas Observatory, in the Atacama Desert high in the Chilean Andes, the telescope had a mirror only 60 centimetres across, tiny compared to the mirrors of most professional telescopes. The resident astronomer didn't even get the chance to do their own observations. Their job was to look after the telescope and observe the list of objects sent by the real astronomers back in Toronto. Most resident astronomers only lasted about a year.

One of the best things about the job, though, was the wonder of the southern sky. It is only in the south that it is possible to see the centre of our galaxy. And it is only in the south that it is possible to see the two closest galaxies. Andromeda may be the nearest big galaxy, but two dwarf galaxies, the Large Magellanic Cloud (LMC) and the Small Magellanic Cloud, are much closer. Like Andromeda, they are visible with the naked eye and are so close that the larger of the two, the LMC, covers 400 times the area of sky of the full moon. These are some of the reasons why Chile, the only country in the southern hemisphere with high mountains, now has more large telescopes than any other country.

The southern sky may be full of wonder, but in the Atacama Desert there are not a lot of things for a young person to do. Ian beat the average and lasted two years as the resident astronomer, but then he left and took jobs back in Canada. But in 1985 he felt the call of the southern sky again and signed back on as the resident astronomer. He spent the next two years at the observatory, working during the night and sleeping during the day. Very occasionally he would go to a bar in a nearby town with some of the other observatory staff.

One of his top targets was the LMC. Back in the 80s, the best way of taking a picture of such a large galaxy was still photography. He spent hours on each exposure, patiently

FIGURE 13.3   A picture in the visible waveband of the Large Magellanic Cloud. Most of the stars in the galaxy are in a bar that stretches from bottom left to top right. The bright red region above the bar is the Tarantula Nebula.

guiding the telescope as he sat in the dome while the emulsion on the photographic plate gradually darkened. Once it finished, he developed the plate in the chemical baths in the telescope's dark room. On these nights, he rarely saw anyone, but he listened to opera, and there was always the company of the telescope itself, the sounds of its motor and gears, as it followed its target across the sky. Ian observed the LMC so many times that he grew to know its geography: its bright diagonal bar of light, its star clusters, the spectacular Tarantula Nebula and the swirls and dapples carved by the dust (Figure 13.3).

On the night of 24 February 1987, he was observing the LMC again. At about four in the morning, he was developing one of the plates in the darkroom. After taking it out of the final chemical bath, he noticed something strange. The galaxy looked different somehow. He thought there might be a bright star where he didn't remember seeing one before. He

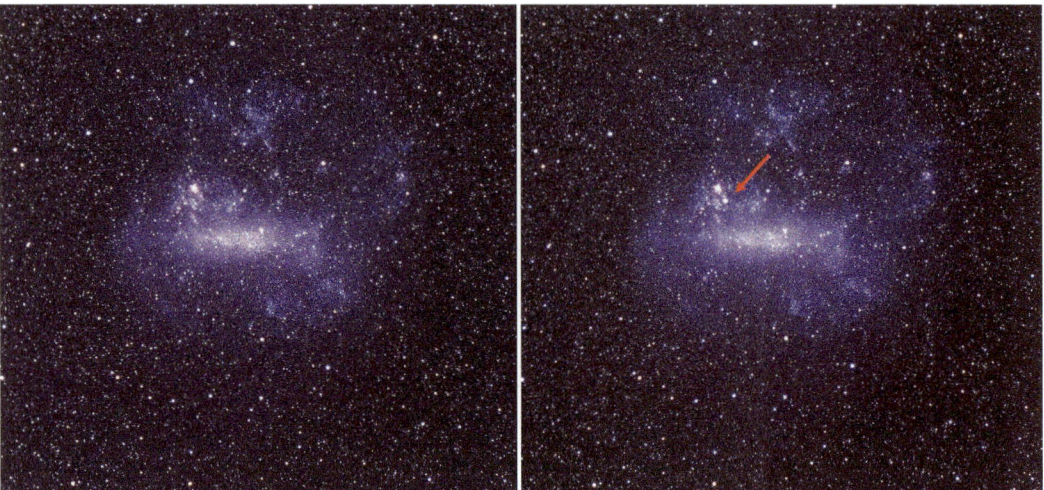

FIGURE 13.4   The Large Magellanic Cloud before (left) and after (right) the explosion of Supernova 1987A. The red arrow shows the supernova.

Credit: ESO

suddenly realised it was so bright he should be able to see it. He put on his coat and went outside.

When he looked towards the LMC, he realised he could see the star just below and to the right of the Tarantula Nebula (Figure 13.4). It was not the brightest star in the sky, but all the other stars in the sky are in the Galaxy. The LMC is another galaxy. He realised that if the star was in the LMC, it must be thousands of times more luminous than every other star in the sky.

He walked over to the dome of the 40-inch telescope to ask the people there what they thought. Everyone went out to take a look. Somebody said it might be a nova, but others said it was so bright it seemed more likely it was a supernova. Someone told Ian that the correct thing to do was to send a telegram to the International Astronomical Union announcing the discovery. After looking up how to do it, Ian sent a telegram announcing the discovery of a possible supernova in the LMC. He had discovered the closest supernova since the invention of the telescope.

Several other people saw Supernova 1987A* that night, but the IAU decided that Ian was the first to see it. (He did go on to become a professional astronomer.)

Although Ian saw it first, it had already been detected, although nobody knew it at the time. Almost 24 hours before Ian spotted the supernova, a pulse of 12 neutrinos was detected by the Kamiokande Neutrino Observatory** in the Kamioka Mine in Japan.

---

* Many supernovae are discovered every year although always in more distant galaxies than the LMC. The naming convention is that the supernova is called after the year it is discovered, with the letter showing the order within the year. Supernova 1987A was therefore the first supernova discovered in 1987.

** A neutrino telescope consists of a large tank of liquid, which is deep underground to shield it from other types of cosmic radiation. Most neutrinos, ghostly particles that flit through planets and stars almost as if they are not there, pass straight through the liquid, but very occasionally a neutrino hits one of its molecules, which emits a flash of light that is detected by one of the detectors arranged around the tank.

It was eventually realised that these 12 neutrinos must have come from the supernova, which, since a pulse of neutrinos is one prediction of the story I told above, shows that it is essentially correct*.

Over the last 36 years, observations of Supernova 1987A have revolutionised just about every aspect of our understanding of supernovae. One of the mysteries they have solved is the origin of interstellar dust.

It is a little surprising this mystery has taken so long to solve because there have only ever been two suspects. One of these has always been supernovae because dust grains are made from the same elements that are made in the core of a massive star, which are ejected by a supernova explosion into interstellar space. The reason they have stayed suspects is that there has been no evidence that any dust is made from these elements before they are ejected into space.

The guilt of the other suspect has been clear for some time. These are the stars with lower masses, ones like the Sun that don't explode as supernovae. Many of the same nuclear fusion reactions occur in these stars, and these stars, too, end their lives by throwing stuff out into space, stuff that among other things forms the beautiful planetary nebulae (Chapter 8). We know that these stars make dust because we have seen it.

The star R Sculptoris is close to the end of its life. The beautiful sub-mm picture of the star taken with the Atacama Large Millimetre Array (Figure 13.5) shows that it is surrounded by rings of dust. As a low-mass star swells into a red supergiant, its outer layers

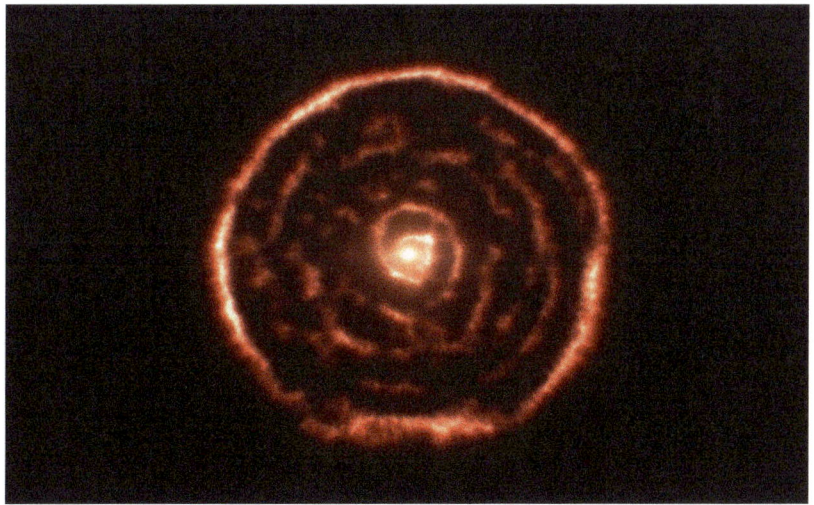

FIGURE 13.5   Sub-mm image taken with ALMA of the star R Sculptoris.

Credit: ALMA (ESO/NAOJ/NRAO)/M. Maercker et al.)

---

* The neutrinos arrived before the light because they hardly interact with matter and passed straight through the outer layers of the star.

become so cold that dust grains start to condense out of the gas. When the star becomes unstable, the dust is blown into space – in almost literally, as the picture shows, a series of smoke rings.

So low-mass stars *do* make dust. The big question has always been whether they are the only source of dust or whether any dust is also manufactured in supernovae. The discovery of Supernova 1987A gave astronomers a great opportunity to answer this question. Unlike my own line of work, in which galaxies change hardly at all in millions of years, in supernova research you get the chance to watch what happens next. The discovery of Supernova 1987A gave astronomers the chance to see if there was any dust formed in the expanding debris. By some estimates, the mass of dust made in a supernova – if supernovae are a significant source of interstellar dust – would be about the same as the mass of the Sun. Even at the distance of the LMC, the infrared radiation from such a large mass of dust would be easy to detect.

But when astronomers looked for infrared radiation from the debris, they detected hardly any. The weak infrared radiation suggested that, at most, there was only a tiny amount of dust, roughly one-thousandth of the Sun's mass, in the debris.[6] Supernova 1987A seemed to have settled the question. Interstellar dust is made in the final stages of the lives of low-mass stars like the Sun. It is not made in supernovae.

Done and dusted then?

Not entirely. By the launch of *Herschel*, 22 years after Ian had spotted the supernova, astronomers had realised there was a problem. By this time, dust had been discovered in galaxies that we are seeing less than one billion years after the big bang, which means there has been only one billion years at the most to make all the dust. This is not a problem if dust is made in supernovae because massive stars have very short lives. But it is a problem if dust is made in low-mass stars. It will be another five billion years before the Sun, for example, reaches the end of its life and ejects any dust into interstellar space. It was not yet a critical problem because there are low-mass stars but with masses a few times greater than the Sun's that have short enough lives to make dust in less than one billion years, but things were beginning to get uncomfortable….

Most astronomers liked maths and physics at school but Mikako Matsuura was not one of them. She did like chemistry and biology, but at school she was mostly interested in sport, especially badminton. It was at home that she became interested in astronomy. Tokyo must be one of the harder places in the world from which to see the night sky, but she remembers when she was about eight being given some binoculars by her father and observing the Moon. Sometime later, after saving up some money, she bought a small telescope and, despite the lights of the city, succeeded in taking pictures of the Moon, Venus and Jupiter.

By the end of high school, she was still not very interested in physics, but she was interested in astronomy. Since she knew a degree in physics was the standard gateway to an astronomy career, she applied to study it at university. In her final school exams, she failed maths and English but still managed to get accepted by Nagoya University, where she knew they did research in astronomy. After her first degree, she went on to do a PhD at a top

Japanese research institution, the Institute of Space and Astronautical Science in Tokyo, and then to a series of postdocs at prestigious institutions around the world.

At launch, Mikako was working as a postdoc in London with Mike Barlow, an expert on the deaths of stars. Mike had suggested that low-mass stars on their deathbeds – swollen unstable stars – and the planetary nebulae they produce would be great first targets for *Herschel*. Mikako was with him when they saw the first spectra, which revealed a multitude of different chemicals, some of which nobody was able to identify (Chapter 8). Huge chemical factories, the stars were a natural research topic for someone who was not much interested in physics and maths but liked chemistry a lot. They also contained a lot of dust, and Mikako gradually became interested in where all the dust had come from.

Mike and Mikako were part of a large international team, led by Margaret Meixner at the Space Telescope Science Institute in Baltimore, that had assembled to apply for time to observe the Large and Small Magellanic Clouds. There are always plenty of reasons to observe these galaxies, and there was no shortage of things to include in the proposal, but since Supernova 1987A happened to have occurred in the LMC, the team thought they should at least mention it.

The one genuine supernova expert in the team was Eli Dwek, a scientist at the NASA Goddard Space Science Centre in the US who has been working on dust and supernovae for over 30 years. By this time, it had been over two decades since the explosion. By 2009 the debris had expanded so much that in *Hubble* pictures (Figure 13.6) it appeared as a faint fuzzy patch at the centre of the spectacular 'ring of pearls', which has nothing to do with the supernova but is material that streamed away from the massive star before the collapse. Eli knew from the weak infrared detections that there must be some dust among the

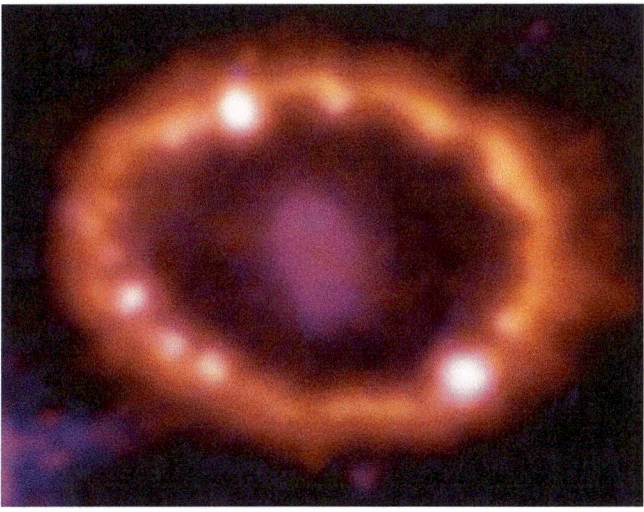

FIGURE 13.6 *Hubble* picture of Supernova 1987A. The 'ring of pearls' is material that streamed away from the massive star before its collapse. The faint patch at its centre is the debris from the supernova.

Credit: ESA, NASA, P.Challis, R. Kirshner (Harvard-Smithsonian Centre for Astrophysics) and B. Sugarman (STScI)

debris. But from the strength of these detections, he calculated that this dust would emit only a tiny amount of sub-mm radiation. He didn't think there was any chance they would detect it with *Herschel*.

Mikako had never worked on supernovae herself, but she thought, just maybe, Eli was being too conservative. Suppose there was some very cold dust among the debris. It might be too cold to emit much infrared radiation, which would explain why it had not been detected before. But cold dust emits its strongest radiation in the sub-mm waveband and *Herschel* was a sub-mm telescope. There was no reason to think this cold dust existed, but it *might* exist. It seemed worth mentioning in the proposal that *Herschel* might detect something, although only if there was some very cold dust among the debris, which nobody thought was likely. Mikako wrote a few lines for the proposal and then forgot about it.

One year after launch, *Herschel* observed the LMC. When the images appeared on the screen, the team were the first to see what the LMC looked like in the sub-mm waveband. They combined the images taken at different wavelengths, as we all did at the time, into a single picture showing the temperature of the dust (Figure 13.7). As with Andromeda,

FIGURE 13.7 The *Herschel* picture of the LMC. It has been made by combining the images at different wavelengths with the colour showing the temperature of the dust: from white and blue (warm dust) to brown (cold dust).

Credit: ESA/NASA/JPL-Caltech/STSci and ESO

the pictures of the LMC in the visible and sub-mm wavebands almost look like pictures of different galaxies. The stellar bar, so obvious in the picture in the visible waveband, has vanished completely in the sub-mm picture.

There is usually a lot of competition in a team for all the most exciting jobs. When it came to checking whether *Herschel* had detected the supernova debris, nobody was really bothered because nobody expected it to be detected. Since Mikako had written the lines about Supernova 1987A in the proposal, Margaret Meixner asked her to do it.

When she looked at the picture, it didn't look hopeful, there was so much sub-mm radiation from everywhere in the galaxy. But when she looked at the exact position of the supernova, she realised she could see a small blob (Figure 13.8). She still wasn't sure whether the blob was merely part of one of the filaments she could see everywhere in the picture. When she calculated the amount of dust needed to emit this much radiation, she was even less sure, her estimate was so high. She thought she must have made a mistake and asked two other members of the team to check her analysis.

They both got the same results. After they had done every check they could think of, the team decided that they really had detected sub-mm radiation from the debris of Supernova 1987A. They calculated that the mass of dust in the debris was about one solar mass – a Sun's worth of dust.

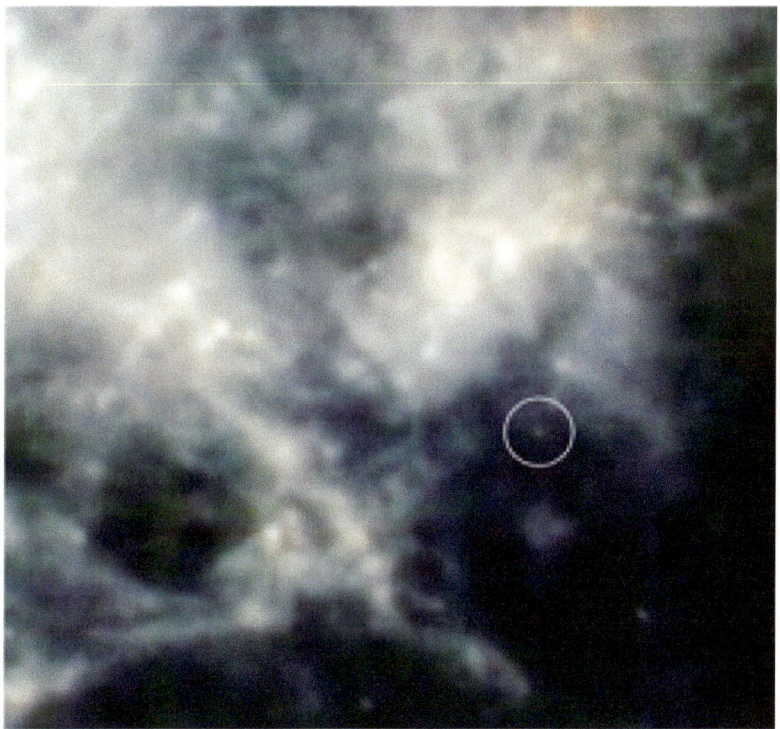

FIGURE 13.8   The sub-mm blob at the position of Supernova 1987A.

Credit: ESA

The dust among the debris is too cold to emit much infrared radiation, which is why it wasn't discovered before. There is no reason to think that Supernova 1987A is special, so if every supernova manufactures a similar amount of dust, this is enough to explain most of the interstellar dust, especially as some is also made in low-mass stars.

Margaret decided that this discovery was important enough to send a paper to *Science*, one of the journals that scientists always use to announce the big discoveries. She asked Mikako to write the paper since she had done all the work. At the end of her paper,[7] Mikako wrote that this discovery explained the existence of dust only shortly after the big bang. As a postdoc on a temporary contract, Mikako must now have thought she had it made. She had made a big discovery, she had just had a paper accepted in a top science journal – surely someone would now offer her a permanent job.

But when she gave a talk about the discovery a few months later at a conference in Stockholm, she discovered that nobody believed her. The problem was the *Herschel* blobs. Everyone pointed out that the *Herschel* picture didn't show enough detail to be sure the sub-mm radiation really was from the debris. There was so little detail, the sub-mm radiation might be from the ring of pearls in the *Hubble* picture, implying it was material that had streamed away from the star before the collapse, or it might be dust that was already there in interstellar space, which had been swept into a pile by the blast wave of the supernova. There wasn't enough detail to tell.

Mikako, as we had been with SDP 81, was now very lucky. Just at the moment she really needed one, a sub-mm telescope that does take pictures that show detail became available. The Atacama Large Millimetre Array was still under construction, but 20 of the 64 dishes were already available for observations. Mikako and her team put in a proposal and were awarded some time.

After ALMA finished its observations, Mikako and her team reduced the data, producing a sub-mm image of the region around the supernova. They combined the ALMA image with a *Hubble* image and an X-ray image, a useful technique that gives a holistic view of an object (Figure 13.9). The visible radiation shows the ring, the X-ray radiation shows the very hot gas in the debris, and the sub-mm radiation reveals any dust that is present. The red in the combined picture shows the dust. The dust is not in the ring – material that streamed off the star before it exploded; it is not at the position of the blast wave – dust that was already present in interstellar space and has been swept up into a pile by the explosion; it is at the centre of the ring among the debris left after the explosion. It is where it should be if it was made in the supernova.

When astronomers saw this picture, they finally believed that supernovae manufacture dust (and Mikako got her permanent job).

Done and dusted.

While Mikako was discovering the origin of interstellar dust, Loretta's and my team were pouring over catalogues and staring at big galaxies on a screen. By late 2011, we had been working on our survey for two years.

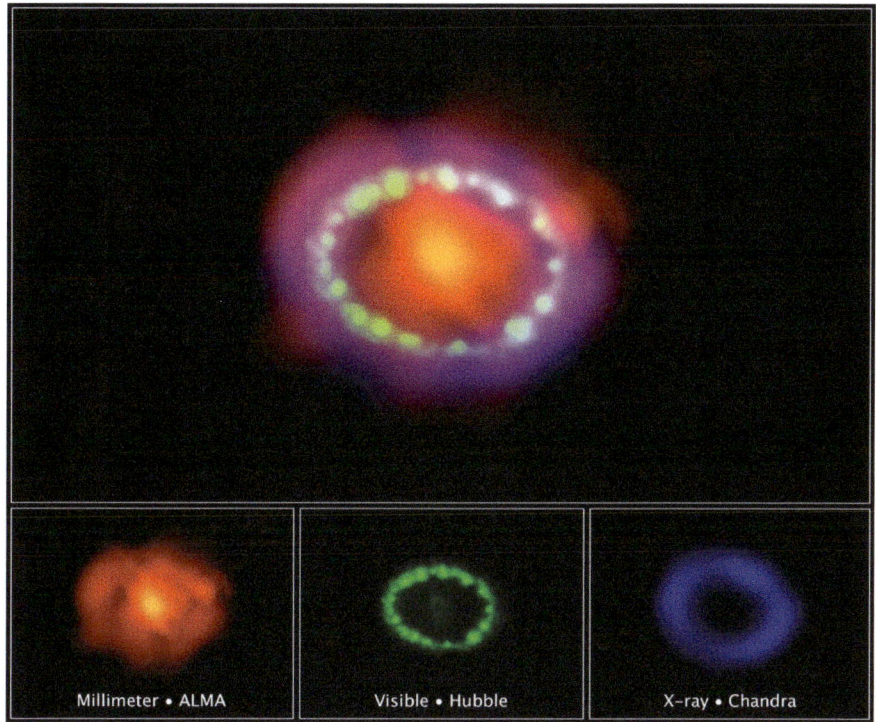

Millimeter • ALMA   Visible • Hubble   X-ray • Chandra

FIGURE 13.9   Picture of the region around Supernova 1987A made by combining three separate images. The blue shows the X-rays, the green shows the visible light and the red shows the sub-mm radiation detected with ALMA.

Credit: ALMA (ESO/NAOJ/NRAO)/A. Angelich. Visible light (NASA/ESA Hubble Space Telescope). X-ray image (The NASA Chandra X-ray Observatory).

When projects last this long there are always changes in the team. Ali Dariush left my group for a job in Cambridge developing methods for analysing medical images; Robbie Auld left astronomy and became an analyst of energy markets; Simon Dye left for a permanent job as a lecturer in Nottingham. Two years after launch, the money to hire postdocs to work on *Herschel* data was beginning to dry up. My group lost three postdocs. I was only able to hire one.

Elisabetta Valiante, like Mattia Negrello, grew up in a small town in Italy, in her case Notaresco, a town of about 6,000 people. Growing up, she was torn between the lure of science and the idea of helping people by becoming a doctor. She decided to do a degree in physics in Rome. She then did a PhD in Munich, the first time she had left Italy, and after that, she travelled halfway around the world to do a postdoc in Vancouver. She joined us in Cardiff to work on the survey in September 2011.

Two months after she arrived, there was a massive upheaval in the team. Loretta and Steve, who had been there from the beginning, decided to emigrate. Loretta had dreamt for years of moving to New Zealand, where she thought there would be a better pace of life and she would be able to spend more time on her outdoor interests: hang-gliding and mountain walking. After she married Steve, they spent a sabbatical together in New Zealand, where Steve also fell in love with the country. By this time, they had spent two years working on the survey, on top of all the regular tasks of their academic jobs. Earlier in the year, they had seen an advertisement for a lecturing job at the University of Canterbury in Christchurch. They put in a joint application asking if they could split the job between them. They were offered the job.

After Loretta told me about the New Zealand job, we agreed that although she and Steve would have less time for research, they would stay members of the team, and Loretta and I would continue to jointly lead the survey. We had a final meeting of the team in Nottingham, and Loretta and Steve flew off to their new life in New Zealand.

CHAPTER 14

The Water Trail

NOTE TO ANYONE SETTING up a big international research project: don't put half of the team in the UK and half in New Zealand. Once Loretta and Steve moved to New Zealand, communication became a big problem. In their new job, they had to do a lot more teaching, which meant less time for research and very little at all for working on a data release, and the people who did have time to do this – me and my two postdocs, Elisabetta Valiante and Matt Smith[*] – were now on the other side of the world. But we couldn't simply get on with it ourselves and ignore Loretta and Steve because they had much of the knowledge we needed. Steve had written the program we were still using to find the sources, and they had been the ones who had devised the ingenious method for identifying the sub-mm-emitting galaxies. Every time we found a bug or something we didn't understand in the program, we needed to talk to Steve. The only way we could make progress, in that pre-pandemic era, was by Skype calls, which were always because of the time difference early in the morning or late in the evening, which meant that somebody or other was often half asleep.

Over the next few years, we got into an annual routine. For most of the year, Elisabetta, Matt and I would work by ourselves in Cardiff on the data release, with regular Skype calls with Loretta and Steve. Then in the summer Loretta and Steve would come over to Cardiff for discussions for a few weeks. Loretta became frustrated that our progress in Cardiff was so slow. It also didn't help matters that my group had decided to make some changes to Steve and Loretta's original data-reduction plan. I thought they were improvements; Steve and Loretta weren't convinced. By the end of their annual visit, after intense discussions, we would just about get ourselves back on the same page. Loretta and Steve would fly back to New Zealand, and then the same thing would start again. By this stage, of course, any deadline for the data release had flown out of the window.

The one positive thing from this period was that we started to put right the big mistake we had made earlier in the project. Loretta was in Sydney Airport on one of her annual trips to the UK. With a lot of time to kill, she started looking again at the pictures in the visible waveband of the galaxies we had found close to the sub-mm sources.

---

[*] I had hired Matt as a postdoc after he finished his PhD.

DOI: 10.1201/9781003195290-14

FIGURE 14.1   Picture in the visible waveband of one of the galaxies close to the position of one of our sub-mm sources.

Credit: *Herschel*-ATLAS team

We had all seen these pictures before. When we looked at them early in the project, we had seen the same kinds of galaxies found in surveys in the visible waveband. There was the occasional red elliptical galaxy – galaxies we quickly realised must be gravitational lenses – but most of them were blue galaxies. These didn't seem any different from the spirals and irregulars we were used to seeing.

But now that she was looking at them closely, Loretta thought they didn't look much like this. Some of the galaxies we had casually assumed were spirals, now that she really looked at them, didn't have clear spiral arms. They appeared diffuse, blotchy – 'little messy dwarfy things', she told me later (Figure 14.1).

So that was something. But apart from that, our team wasn't doing anything new. It was just the same old business of slowly improving our sub-mm images and catalogues so that we could eventually release them to the community, which now seemed further off than ever.

Everyone else, though, did seem to be still having fun. Every week there seemed to be an announcement of some new big discovery. Occasionally there was something completely unexpected. At one of the *Herschel* conferences, somebody showed a spectrum of some supernova debris with a spectral line he said nobody had been able to identify. Then someone at the back of the auditorium shouted out that it was argonium, which is a chemical compound of hydrogen and argon. And if that doesn't sound exciting, argon is one of the 'noble gases', which are not supposed to chemically combine with anything else. Except in space they do!

Some of the discoveries were connected to life itself. Of all the origin questions, the origin of life is surely the big one – that and all the questions that follow: Is there life elsewhere in the universe? Did life start on this planet or did it come from somewhere else? Is the evolution of intelligent life inevitable? And so on. The one thing I regret about my career is I have never done this kind of research myself. Anyone whose original motive for becoming an astronomer was watching Doctor Who or Star Trek probably has similar feelings, but back in the 80s when I started out all these questions seemed too remote. There were only a few universities where scientists were even thinking about them.

One of the big challenges in thinking about these questions is that so far we have only one example of life, and even here on Earth its beginnings are hidden by four billion years of geological upheaval, which mangles the fossil record, and of evolution, which efficiently wipes out previous versions of life. The other problem, which is more fundamental, is that with only one example we have no way to tell whether life is incredibly rare or whether the Galaxy is brimming with life. Suppose that life has arisen only one time in the history of the universe – that the Earth is the only one of trillions of planets on which there is life. In this vast cosmic lottery, we and the rest of life on Earth must necessarily have bought the winning ticket. We therefore have no way of knowing whether we are alone in the universe or whether life is as common as odd socks.

It is almost impossible to overstate how ignorant we were back in the 80s. This ignorance was neatly summed up by an equation dreamed up in the 60s by the astronomer Frank Drake, who was using a radio-telescope to search for signals from extra-terrestrial civilisations. He devised his equation to estimate the number of civilisations he might discover. In my update, the number of broadcasting civilisations, N_{ET}, is given by:

$$N_{ET} = N_* \times f_p \times n_e \times f_l \times f_i \times f_c \times L$$

In this equation, N_* is the number of stars in the Galaxy; f_p is the fraction of these with planets; n_e is the average number of planets around each of these stars; f_l is the fraction of these planets on which life begins; f_i is the fraction of these on which intelligent life then develops (bacteria would not be great conversationalists); f_c is the fraction of these lifeforms that develop radio technology; and L is the fraction of the lifetime of the Galaxy during which that civilisation is broadcasting radio signals (If we are really unlucky, there may have been many civilisations in the Galaxy but none of us have been broadcasting at the same time.).

Here is what we knew about these then:

N_* – there are roughly 300 billion stars in the Galaxy. We did know this back in the 80s.

f_p – we obviously knew the Sun has planets but we didn't know then whether any other star has planets. They might all have planets or none might.

n_e – we knew the number of planets around the Sun but obviously not for any other star.

f_l – didn't know this because of the cosmic lottery problem.

f_i – no idea.

f_c – no idea.

L – no idea.

Basically, apart from the first factor, it was all guesswork. If you were feeling optimistic, you might guess values that would make the equation predict there was probably an extra-terrestrial civilisation around every star. If you were feeling pessimistic, you might guess values that would show we are all alone in the universe.

In 1994, everything changed. Two groups – Didier Queloz and Michel Mayor at the University of Geneva and Robert Butler and Geoffrey Marcy at San Francisco State University – had started searching for planets around other stars by looking for changes in the speed of a star caused by the gravity of an orbiting planet.[†] Few people thought they stood much chance. In the solar system, the big planets, Jupiter and Saturn, are so far from the Sun that their gravity doesn't disturb it much – the changes in the Sun's speed are very small. But these scientists were very lucky, or had great intuition, or maybe just the common sense to realise that the solar system may not be the pattern for all planetary systems. As it happens, in other planetary systems the big planets are often very close to their stars, often even closer than Mercury is to the Sun. In these planetary systems, the gravitational force exerted by the big planets makes the star's speed change a lot, much more than anyone had expected. The teams detected oscillations in the speed of the star as the planets orbited around it, and for the first time we had evidence that our star is not the only one with planets.[‡]

But this is all a little abstract. We now knew, because of these osciillations in the speed of the star, that there are planets around other stars, but nobody had seen one of these planetary systems. Arguably, the first person who saw one was a sub-mm astronomer.

Jane Greaves is one of the most original scientists I have ever met. It may be surprising to anyone who is not a scientist that it is perfectly possible to be a good scientist without ever having an original thought. When a new telescope becomes available, for example, it's usually obvious what are the good things to do (carry out a survey of a large area of sky!). Jane, though, is someone who has rarely done the obvious. Lured like many of us into astronomy by science fiction, for her it was a book: *A Fall of Moondust* by Arthur C. Clarke. This book is about an accident to a moon-bus full of tourists visiting the Moon. It describes how the bus, lost beneath the lunar dust, is eventually discovered with an infrared camera operated by a grumpy infrared astronomer – which is what Jane claims she has now turned into.

She was in the news a couple of years ago when her team discovered phosphine in the atmosphere of Venus. Phosphine is a biomarker, a sign there might be life on a planet. Venus is often called the Hell Planet, a planet on which, if you were stupid enough to land on its surface, you would be immediately killed in four different ways,[§] so searching for life there is about as far from obvious as you can possibly get.

[†] They did this by taking repeated spectra of a star, looking for small changes in the wavelengths of its spectral lines that are the result of oscillations in the star's speed caused by the gravitational effect of a planet.

[‡] Michel Mayor and Didier Queloz won the Nobel Prize in Physics in 2019 for this discovery, which they shared with the cosmologist Jim Peebles.

[§] Broiled, crushed, asphyxiated, poisoned.

It was Carl Sagan, another out-of-the-box scientist, who was the first to suggest there might be life there. He argued that despite the deadliness of the surface, it would still be possible for life to exist in the cloud layers,[1] with one suggestion being organisms that float in the atmosphere using hydrogen-filled bladders.[2] For almost 60 years, nobody bothered to follow up this suggestion, which seemed like a wacky idea from science fiction, until Jane came along. She decided it was worth having a look. She decided to look for a biomarker, a molecule whose main source is biological-like the oxygen in our atmosphere, which if all life suddenly vanished would gradually disappear.[*] She chose phosphine.

She applied for observing time on the Atacama Large Millimeter Array (ALMA) to look for the phosphine, choosing to apply through Director's Discretionary Time, the route meant for urgent scientific projects. The director, clearly thinking that testing ideas from 60 years ago was not very urgent, turned her down, but she persevered. She and her team were eventually awarded observing time with both ALMA and the James Clerk Maxwell Telescope to look for the phosphine.

They did find evidence there is phosphine in Venus' atmosphere, evidence for Sagan's idea that there is life high in the atmosphere. It did not prove there is life there: phosphine can also be produced by natural geological and chemical processes, although Jane's team do not think these processes on Venus can explain the amount of phosphine they found.[3] So there may be life on the Hell Planet, or maybe once we learn more about the geological and chemical processes there, we will find some other explanation of the phosphine.[**]

Back in the 90s, though, Jane was not a professor but a young scientist just starting out. She was working at the Royal Observatory Edinburgh, where the world's first sub-mm camera, SCUBA, was built. SCUBA had recently been shipped out to Hawaii and, as the first-ever camera in a new waveband, everyone wanted to use it (Chapter 4). Jane's plan was that she would use it to solve a mystery: stars that emit intense sub-mm and far-infrared radiation.

These had been discovered in the 80s with the Infrared Astronomy Satellite (IRAS). Any telescope needs calibration stars, ones whose brightness is known very precisely. Vega, one of the brightest stars in the sky, is one of the standard calibration stars everyone uses. But when the IRAS scientists observed Vega for the first time after launch, they discovered that it was emitting far more far-infrared radiation than expected. They soon discovered other stars like this, stars that for some reason are strong sources of far-infrared and sub-mm radiation.[††]

Nobody knew what was emitting this radiation. Was it the star itself or was it being emitted by something around the star? The obvious thing to do in a situation like this is to take a picture that shows enough detail to reveal the source of the radiation. IRAS had

---

[*] The oxygen, which is highly reactive, would gradually combine with other gases in the atmosphere and rocks on the surface.

[**] This was the argument made by Jane and her team. They didn't claim their detection of phosphine proved there was life on Venus, acknowledging the possibility of some undiscovered non-biological process that might explain the phosphine.[3]

[††] IRAS detected far-infrared radiation, but the sub-mm waveband is the next one over. If a star is a strong far-infrared source, it is likely to be a strong sub-mm source.

shown there was far-infrared radiation from the general vicinity of the star, but it didn't have a camera. Back in the 80s that was the big problem. There were no cameras that took pictures in either the far-infrared or sub-mm wavebands.

SCUBA changed everything. When SCUBA became available, Jane put in a proposal to use it to take sub-mm pictures of the mysterious stars. This was one of her more obvious ideas and the proposal was accepted.

A short while later, in August 1997, Jane and her collaborator Wayne Holland were in the control room of the James Clerk Maxwell Telescope ready to start their observing run. I imagine they were tense. The first night of an observing run is usually tense, and it's worse when the instrument is a new one, especially one that everyone expects to make big discoveries.

As the sun went down, they observed the first of the stars on their list. After the exposure ended, they began an observation of a second star. While the telescope observed their second target, they reduced the data from the first observation. When they looked at the picture of the star on the screen, all they could see was noise.

The second star was Epsilon Eridani. Epsilon Eridani is one of the closest stars, only about ten light-years away. It is also very similar to the Sun. When they finished the exposure, they ran the sequence of data-reduction programs and put the picture up on the screen.

The telescope operator came rushing over. Jane remembers thinking, *I think we've found another solar system*. By this time, astronomers knew there were other planetary systems, but it was all indirect, abstract. Nobody had *seen* a planetary system. This was a picture of one.

The picture, strangely, is of something that is impossible to see in our own planetary system (Figure 14.2). The picture does not show planets but, as it always does with sub-mm pictures, tiny solid particles – dust. The dust is at about the same distance from Epsilon Eridani as the belt of small objects beyond the orbit of Neptune, the Edgeworth-Kuiper Belt, is from the Sun.[‡‡] We now think that the source of the far-infrared and sub-mm radiation from the mysterious stars is debris, the host of tiny particles produced by collisions between the small objects in the belts around the stars.[4] We can't see the radiation from this debris in our own planetary system so clearly because we are surrounded by it. Jane and Wayne's picture was not a picture of the planets, but it was still a picture of a planetary system.[§§]

Sub-mm pictures like this show that a planetary system is present. They have also begun to reveal things about the formation of a planetary system. Dust, on top of all its other good qualities, gives us a way of observing this as it happens.

As I explained earlier, the birth of a star happens as the result of the collapse of a cloud of gas and dust (Chapter 9). The formation of the planetary system happens at the same time.

The collapsing cloud is likely to be rotating, which means that apart from gravity pulling it inwards there is also an outwards force caused by the rotation (the same force that is felt on a roundabout). The effect of this outwards force is that the cloud collapses at a different

---

[‡‡] We now think that Pluto, rather than being a planet, is one of these objects.

[§§] The first picture of the dust around one of these stars was a picture in the visible waveband of the star Beta Pictoris, which showed starlight reflected by the dust.[5] But the planetary system around Beta Pictoris is seen edge-on, so the picture just shows a streak of light. It has nothing like the visual impact of Jane's picture.

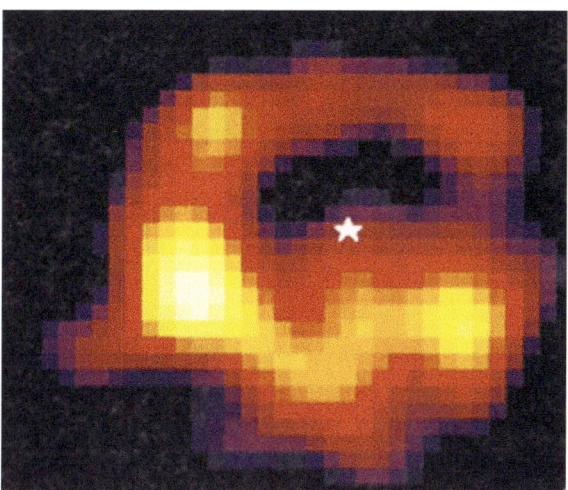

FIGURE 14.2   Sub-mm picture of Epsilon Eridani.[4] The asterisk shows the position of the star, which we can't see because it is the dust around the star that emits the sub-mm radiation rather than the star itself.

Credit: James Clerk Maxwell Telescope, Jane Greaves

rate in different directions, with the result that much of the gas ends up in a disk that is rotating around the proto-star. As I explained in Chapter 9, a consequence of the collapse is that the gas becomes very hot, volatilising any solid material in the disk. Once the disk has stabilised, though, it starts to cool and solid particles begin to condense out of the gas.

Much of what happens next is very uncertain. The solid particles – dust grains – start to clump together. Eventually, there comes a time in which there are billions of solid objects within the disk, each between ten and a few hundred kilometres in size.

After another period of coalescence, many of these objects are incorporated in the planets, with the rest either being slung into interstellar space by the gravity of the newly formed planets or surviving in belts of small objects between the planets. If this story is correct, the two belts of small objects in the solar system are relics of this process.

The outlines of this story were first proposed by the Marquis de Laplace 200 years ago, but, as we often say, it is a bit 'hand wavy'. There are not many details. To provide some details, it would really help if we could observe the birth of a planetary system at all its different stages. Even with the biggest telescopes on Earth, it's not yet possible to take pictures of the planets around other stars", but dust, surprisingly, gives us a way of observing what happens at the time of the birth of the planets.

This is possible because of dust's amazing ability to emit radiation. The sub-mm radiation from a planet around another star is much too faint to detect, let alone one of the asteroid-sized objects that were there before the planets. But imagine that one of these smaller

---

" The Extremely Large Telescope, currently under construction in Chile, will be able to take pictures of planets around other stars.

objects is blown by the phasors of some interstellar spaceship into a cloud of tiny particles each about the size of an interstellar dust grain. The combined surface area of all the particles will be roughly ten billion times that of the original object,*** and because it is the surface area of an object that determines how much radiation it emits, the sub-mm radiation from the cloud of dust will be about ten billion times stronger than the sub-mm radiation from the original object. It is this marvellous property of dust that makes sub-mm observations the best way we have of observing the birth of a planetary system.

Epsilon Eridani is well past the planet-forming stage, but the star HL Tau is less than 100,000 years old.

The sub-mm picture of HL Tau made with ALMA shows the birth of a planetary system while it is happening (Figure 14.3). There are no planets in the picture, or the star itself, because these emit hardly any sub-mm radiation, much too little to be detected with ALMA. The stuff that is visible is the dust, the solid particles that have condensed within the gas. There are no planets in the picture, but the rings are a sign they may be there

FIGURE 14.3 Sub-mm image of the disk around HL Tau made with the Atacama Large Millimetre Array.

Credit: ALMA (ESO/NAOJ/NRAO)

*** The reason for this is that the volume of a sphere is proportional to the cube of its radius, but the area of its surface is proportional to the square of its radius. So if you divide a sphere into 1,000 equal-sized smaller spheres, each sphere will have a radius one tenth that of the original sphere and the area of its surface will be one hundred times smaller. But there are now 1,000 spheres, so the combined area of all the spheres is 10 times greater than that of the original sphere.

FIGURE 14.4 *Herschel* image of Fomalhaut. The star is at the position of the blob in the centre, although the radiation there is probably not from the star but from hot dust or gas close to the star.[6]

Credit: ESA

already. There are other possible explanations but the most likely is that the gaps between the rings have been swept clear of dust by orbiting planets – invisible cosmic bulldozers.

After *Herschel's* launch, the disks around young stars like the one around HL Tau and the debris that had been discovered around older stars like Epsilon Eridani both became obvious targets. Jane Greaves was part of a team that was awarded time to observe the debris stars.

One of their top targets was Fomalhaut, one of the brightest stars in the southern sky. When Jane first saw the *Herschel* picture of Fomalhaut (Figure 14.4), which was taken with PACS, it was so perfect she thought she must be seeing a fake image rather than a real one. The ring in the picture is uneven, brighter at the top and bottom and not exactly centred on the star, which again is almost certainly caused by disturbance of the dust by the unseen planets.[6] The team was able, for the first time, to find out what the dust is made of.

Before launch, we did know that interstellar dust is mostly made from silicates. Silicates are made from a network of chemical pyramids, with oxygen atoms at the four corners of each pyramid and a silicon atom, the buried pharaoh, at its centre (Figure 14.5). The pyramids are arranged in different ways in different types of silicate: in chains, sheets and sometimes in a three-dimensional lattice. The silicates are important: 95% of the Earth's crust and 97% of its mantle are made from them, so it is perhaps not surprising that they are also important in interstellar dust, although at the time of launch we didn't know what kind of silicate.

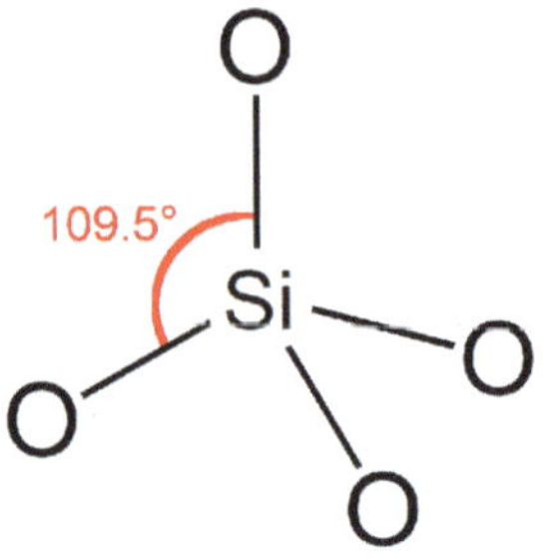

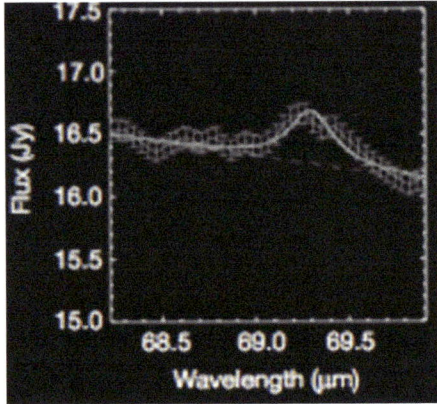

FIGURE 14.5   The silicate pyramid, with oxygen atoms at the corners of the pyramid and a silicon atom at the centre.

FIGURE 14.6   *Herschel* spectrum of Beta Pictoris.[7]

Credit: ESA

The team used the spectrometer part of PACS to investigate the composition of the dust. The pinnacle in their career as interstellar geologists came when they obtained a spectrum of the dust debris around the star Beta Pictoris (Figure 14.6). Unlike the narrow spectral lines in optical spectra, the spectrum of a lump of rock generally shows only broad bumps, but the principle is the same: the wavelengths of the bumps tell us which minerals are present in the rock.[†††] There was only a single bump in the spectrum of Beta Pictoris, but the team used its wavelength to deduce the mineralogical composition of the dust. They discovered that the dust around Beta Pictoris is not just made of silicates. It is made of one particular kind of silicate: olivine.

But that was only the first step. The pyramids are linked together in olivine, as they are in many silicates, by metal atoms – magnesium and iron in the case of olivine. The precise wavelength of the spectral bump depends on the exact proportions of the two elements. By carefully measuring its wavelength, the team discovered that the pyramids in the dust around Beta Pictoris are held together almost entirely by magnesium atoms. By

---

[†††] Minerals generally produce broad spectral 'bumps' rather than narrow spectral lines because they are chemically more complex than the chemicals that produce the spectral lines in an optical spectrum.

using *Herschel* to observe the dust around a star 60 light-years away, they were able not only to discover its mineralogical makeup but also its exact chemical composition.

Olivine also happens to be the silicate that makes up most of the Earth's upper mantle, so the dust around Beta Pictoris, a star 60 light-years away, is largely made from one of the most important minerals on our own planet. On our planet there are exactly four beaches on which the sand is green[‡‡‡]. Green sand is made from the same magnesium-rich olivine. So if you ever get the chance to walk on one, you can tell yourself you are burying your toes in interstellar dust.

The planet-forming disks, like the one around HL Tau, were also prime targets for *Herschel*. But in the case of the disks, it was not the dust that was the lure for the *Herschel* teams. It was something chemically simpler and far more familiar: water.

At home on my desk I have the fossil of a double ammonite that I found in an antique shop (Figure 14.7). I can reach out and touch the fossils of two creatures who, 200 million years ago, were swimming around in the oceans of the Jurassic period. It has been the

FIGURE 14.7   The fossil of two ammonites.

Credit: ESA

[‡‡‡] Papakolea Beach in Hawaii, Talofofo Beach in Guam, Punta Cormorant Beach in Ecuador and the shore of Lake Hornindalsvatnet in Norway.

oceans that have been the venue for 90% of the history of life on Earth. It is only comparatively recently, only about 400 million years ago, that life began to colonise the land. Our own family tree leads back to the amniotes, who left the oceans after the evolution of the amniotic membranes that make it possible for human embryos today to develop in the salty chemistry of the sea.[8]

Water is also almost certainly where life began. Back in the 50s, the chemists Stanley Miller and Harold Urey carried out a beautiful experiment. They filled a sealed glass vessel with the gases they thought must have formed the early atmosphere of the Earth, with a little water at the bottom of the vessel. Then, to simulate lightning in the atmosphere, they set up an electrical circuit that sent electrical sparks through the gases. When they came back a week later, the water had turned red. When the water was analysed, they discovered it now contained amino acids, the building blocks of the proteins, one of the two most important constituents[§§§] of life on Earth. Their experiment showed that given the right raw ingredients, with a dash of energy, life can be whipped up - although whether this is something infinitesimally rare or as common as a nice Victoria sponge is still the big question. There are many locations that have been suggested for the origin moment, from small ponds to the hydrothermal vents at mid-ocean ridges,[8] but the thing that is always common is water.

Water is a unique chemical because of the properties of its molecules. I have always wondered whether the director of the movie *The Shape of Water* gave it that name because of the shape of the water molecule. The abrupt bend between the two hydrogen-oxygen bonds[¶¶¶] and the affinity of the oxygen atom for electrons makes water perfect for dissolving chemicals. It is often called the universal solvent, and it is probably not surprising that the chemical reactions in our bodies all occur in solution in water. A not-very-challenging question for a pub quiz is the fraction of our body weight that is water. The answer according to my sources on the internet is between 60 and 70%. Water is so unique that it is difficult to think of any other chemical that could take its place.

(Although it is always possible we are biased. We are largely made of water, so we may be looking at life through prejudiced eyes – our eyes are little bags of water, after all.)

Water is also something that makes life on Earth a little safer. Harold Urey also discovered the carbonate-silicate cycle. Rainwater is slightly acidic because it contains dissolved carbon dioxide from the atmosphere. Over long periods of time, this slightly acidic water dissolves the silicate rocks, with the rivers carrying the products down to the ocean, where they eventually become incorporated into the shells of sea creatures. When these die, they sink to the bottom of the ocean where their compressed remains form sedimentary carbonate rocks like limestone and chalk. This process takes carbon dioxide out of the atmosphere. The other side of the cycle is when a tectonic plate is pushed down into the mantle, where any carbonate rocks in the plate are melted and the liberated carbon dioxide released back into the atmosphere through volcanos.

[§§§] The other are the nucleic acids.
[¶¶¶] 104.5 degrees.

This cycle is our planet's thermostat. Any increase in the atmospheric carbon dioxide increases the greenhouse effect, and the increased temperature leads to higher rainfall because of greater evaporation from the oceans, which increases the silicate weathering and the rate at which carbon dioxide is taken out of the atmosphere. Any fall in atmospheric carbon dioxide, on the other hand, leads to lower rainfall and a reduction in silicate weathering, while leaving the rate unchanged at which carbon dioxide is pumped back into the atmosphere through volcanos – so the level begins to climb again. This feedback process resembles many in the human body**** and acts to stabilise the amount of carbon dioxide in the atmosphere. (It can't cope unfortunately with the rate at which we are currently releasing carbon dioxide into the atmosphere.)

Water's centrality to life – although we *are* looking at it through water-filled eyes – has often led to the assumption that the only place in the solar system suitable for life is our planet. It is true that the Earth is the only planet currently in the region in which water can exist in its liquid form – the 'habitable zone', also sometimes called the 'Goldilocks Zone' because the temperature there is neither too hot nor too cold.

But we now know that the concept of a habitable zone is too simplistic. For a start, the position of the habitable zone is not fixed because of changes in the Sun's energy output and in the planets themselves.†††† Even though Mars may not seem suitable for life now, there are features on its surface that show there was once running water, which shows it must once have had a dense atmosphere.‡‡‡‡ And if life did get started on Mars billions of years ago, it might still exist today; some of the extremophiles on Earth survive in worse environments. It is even possible that there was once water running over the surface of Venus,[9] which might explain any organisms high in the atmosphere, if life started on the surface but migrated upwards when conditions there became too bad.

We also now know there are places well outside the habitable zone where there is liquid water. Most of the water in the solar system is not in the Earth's oceans and icecaps but in the asteroids, comets and the moons of the giant planets.[9] These are outside the habitable zone, so most of this water is locked up in ice, but on some of the moons – Europa (Figure 14.8), Ganymede and Callisto around Jupiter and Enceladus around Saturn – there is now evidence there are oceans under the ice. These are kept liquid by the gravitational field of the giant planet, which generates heat within the moon as it orbits the planet, clenching and unclenching the moon like a tennis ball. These oceans under the ice have risen to the top of the list of places to look for life (along with the surface of Mars), which

---

**** It is feedback processes like this that led the chemist James Lovelock and the microbiologist Lynn Margulis to propose the Gaia hypothesis, that the Earth seems in some ways to behave like a living creature with a self-regulating metabolism.

†††† The temperature of a planet is partly set by the energy output of the Sun and partly by whether there is a greenhouse effect on the planet, which depends on the composition of its atmosphere. If there were no greenhouse effect from the Earth's atmosphere, for example, the oceans would be frozen.

‡‡‡‡ There are features on the surface that look like riverbeds, gullies cut by flash floods and even possibly the shore of an ocean. During its first billion years, the atmosphere must have been dense enough to sustain a sizable greenhouse effect, making possible rivers and possibly an ocean. As the planet gradually lost its atmosphere, the temperature dropped. The water is still there, but it is now locked up in ice.

FIGURE 14.8   the icy surface of Europa. There is an ocean under the surface, kept liquid by the heat released by the kneading of the moon by Jupiter's titanic gravitational field. The cracks in the ice caused by the kneading can be seen in the picture.

Credit: NASA/JPL-Caltech/SETI Institute

was one of the motivations for the Jupiter Icy Moons Explorer (JUICE) launched by ESA in Summer 2023 and currently travelling towards Jupiter and its moons.

Everywhere we look now in the solar system there seems to be water. Even the Moon, once thought to be bone dry, is now known to contain deposits of ice. If life does need water, there is a lot of it around.

But what is the source of the water? The water molecule is very stable. Any water molecule on earth today has been through incessant transformation – sometimes in an ocean, sometimes in a raindrop, sometimes in a snowflake or an ice particle in a cirrus cloud – but the molecule itself always remains the same. The water molecules I see coming out of my tap are the same water molecules that were there on Earth four-and-a-half billion years ago. But where did they all come from before that?

And that's where *Herschel* came in.

The big problem with water, if you are an astronomer, is that we have a little too much of it ourselves. We live on the blue planet, which is usually a good thing, but it also means the atmosphere is full of water vapour. The spectral lines from the water in giant molecular clouds are very bright and should be easy to detect, but almost all this radiation is absorbed by the water in our atmosphere before it is able to reach a telescope. The disadvantage of living on such a watery world is that the only way to observe the water in the rest of the universe is to go into space.

Which, of course, is why *Herschel* was built in the first place. The beautiful pictures of dust, all the discoveries about gravitational lenses, planetary nebulae, supernovae, the origins of stars and galaxies, and my own survey – none of this was the original reason for building *Herschel*. That all came later. Back in the 80s, the original motive for *Herschel*, or FIRST as it was then called, was to send a telescope into space, above our water-soaked atmosphere, to look for the water in the rest of the universe (Chapter 5).

HIFI, *Herschel*'s third instrument, was the instrument designed to look for the water, which is why the day it died was such a bad one. The engineers now thought they knew what had gone wrong (Chapter 10). During one of the contact periods, they reprogrammed HIFI so that it used the spare electronics and so the guilty switch stayed shut during a reboot. They thought the diode should not burn out now. Everyone crossed their fingers. There was only one set of spare electronics.

Six months after HIFI died, somebody somewhere pressed a button. HIFI woke up. The diode didn't burn out. HIFI was good to go.

The main team of water diviners was the WISH team, which as acronyms go is delightful, although looking at the full name[§§§§] I can't resist saying it's rather wishful. The team's leader was the Dutch astronomer Erwine van Dishoeck, who has had a distinguished career investigating the physics and chemistry of interstellar gas. The main goal of their observing programme was to follow the water along the 'water trail' to find its source.

Before launch, we already knew how most of the water in the universe was made. It was not made in the interstellar gas. The density of the gas is so low, 100 billion billion times less than the density of our atmosphere, that hydrogen and oxygen atoms rarely encounter each other. If the gas was all there was, the universe would be very dry.

One place, though, where hydrogen and oxygen atoms do meet is on the surface of a dust grain. When a gas atom hits a dust grain, it will often stick loosely to it, which means the atom can then wander around the surface, encountering other atoms. We were therefore confident that most of the water was made from random encounters between hydrogen and oxygen atoms on the surfaces of dust grains – dating sites for atoms (which adds another to the list of reasons why dust is important). We also knew that the main reservoirs of water are the giant molecular clouds, simply because that's where most of the gas and dust is.

The big question, though, was whether these reservoirs were the ultimate source of the water in the Earth's oceans and that flows out of my tap. Nobody was sure of the answer because of all the things that happen to a water molecule along its way from the giant molecular cloud.

The embarkation point for a water molecule is a dense region within one of these clouds – a molecular core, a particularly dense region that will eventually collapse to form a star.

---

§§§§ Water in Star Forming Regions with the Herschel Space Observatory – try to pick out the acronym.

Molecular cores are very cold, so as it starts out on the water trail, the water molecule is embedded in a layer of ice surrounding one of the dust grains in the core.

When the molecular core collapses, the water molecule may end up in the proto-star, in which case we say goodbye to it. One of the questions at launch was whether, if instead the water molecule ended up in the disk surrounding the proto-star, would it be able to survive there? Most of the disk is very hot, so it seemed likely, if it ended up in the disk, the ice would be melted and the water molecule would enter the gas phase. But the big question was: Is the temperature so high in the disk that the water molecules there would be split back into hydrogen and oxygen atoms? If this happens, the water that comes out of my tap must have been made in a different way, almost certainly being formed in the disk around the newly formed Sun as the gas in the disk cooled and the hydrogen and oxygen atoms eventually recombined.

If a water molecule that travelled the water trail to the solar system did survive the heat in the disk, it may have ended up incorporated in one of the other planets or one of the small objects that were hurled out of the solar system (and we again say goodbye to it). If it survived all this, there is a chance it will have ended up as one of the molecules in the Earth's oceans, but exactly how this happened was one of the other questions for which we didn't have an answer when *Herschel* was launched.

Once HIFI was working again, the WISH team set off to find out. One of their first targets was an object right at the beginning of the trail, the molecular core L1544.[10] This is a core that has just started to collapse. Before they made their observations, nobody knew how much water is contained in a core like this because almost all of it is locked up in ice, which is almost impossible to detect. The team's plan was to use HIFI to detect the small amount of water vapour that is always present where there is ice – there are always some lively molecules that manage to escape from the ice – and use their measurement of the amount of water vapour to estimate how much ice there must be in the core.

Even with the biggest telescope ever sent into space, the team had to make one of the longest ever *Herschel* exposures to detect the water vapour. They had to observe L1554 for almost 14 hours, but eventually they detected one of the spectral lines of water. They used this detection to estimate that the mass of water vapour in L1554 is about 2000 times the mass of all the water in the Earth's oceans. But there is a lot more water locked up in the ice. They estimated that the mass of all the ice in L1554 is roughly three times the mass of Jupiter, which would be enough to fill the Earth's oceans three million times over.[10]

The amount of water in the universe has always been wildly uncertain because so much is locked up in ice, which is why carbon dioxide is often given in lists as the second most abundant molecule after molecular hydrogen.[11] But the WISH team found that in L1554 there are more molecules of water than of carbon dioxide. So if L1554 is representative of the universe – and there is no reason to think it isn't – water is number two.

To investigate whether water molecules can survive in the heat of the disk around a newly formed star, the team observed the disk around the star TW Hydrae. The disk around this star, like the disk around HL Tau, is one in which planets have either formed already or are about to form.

The team again detected a spectral line from water vapour, which they used to estimate the amount of water vapour in the disk. Again, they were able to use the amount of water vapour to estimate how much ice must exist in the disk. They estimated that the ice in the disk around TW Hydra contains enough water to fill the Earth's oceans several thousand times over.[12] They also showed that most of the ice is in a part of the disk where the temperature has never been high enough to melt the ice, let alone split the water molecules into hydrogen and oxygen atoms.[13]

It therefore seems likely that a water molecule does survive the journey from a giant molecular cloud to the disk around a newly formed star. The remaining question was: what happens next? How does it make the journey from the disk to the surface of a planet? Being a little planetcentric: How did the water in the Earth's oceans get there? Before launch, there were two possible answers.

One possible answer is that the water molecules in the disk around the newly formed Sun ended up incorporated in the Earth itself. Then, as the newly formed planet slowly cooled, the water gradually trickled up to its surface to become the oceans.

The other possible answer is that the water was transported to the Earth during the era of bombardments that followed the formation of the planets. After the formation of the planets, there would still have been many smaller objects, whose relics today are the asteroids and comets, roaming the disk. Most of these would have been slung into interstellar space by the gravity of the planets or ended up in the solar system's two belts of small objects, but there would also have been many that smashed into the planets' surfaces, which we know did happen because of the craters on the Moon. There are large amounts of ice in comets and in some asteroids,[9] and it seemed a reasonable guess that the objects like this that bombarded the young Earth might also have delivered enough water to fill the oceans.

The second answer was probably the more popular one, but at the time of launch the big problem was that it had flunked a very basic test. One way of checking the source of any water is the proportion of 'heavy water', water molecules in which one of the hydrogen atoms contains an extra neutron. The proportion of heavy water is not high, usually about two out of every 10,000 water molecules, but it is fixed. If the water in a local reservoir has a different proportion of heavy water to the water coming out of my tap, that reservoir can't be the source of my water. By launch, the heavy-water proportion had been measured in six different comets.[14] In all of them, it was about twice the value in the oceans. It therefore didn't seem possible that comets could have delivered our water.

All six, however, did happen to belong to just one of the two classes of comet. All six were comets that visit the solar system from interstellar space, pass close to the Sun, and then leave the solar system never to return. The other class are the homebodies. The home of these objects is the Edgeworth-Kuiper Belt beyond the orbit of Neptune, where they stay until they are disturbed by the gravitational force of some nearby object and plummet towards the Sun. They remain in the solar system in orbits that pass close to the Sun until all the ice has melted and all that is left is some orbiting debris.

Very soon after HIFI started working again, a team led by the German astronomer Paul Hartogh used it to measure the proportion of heavy water in one of the homebody comets,

Comet 103P/Hartley 2.[14] They measured a value almost exactly the same as the value in the oceans. It is only one comet, of course, but it is now possible again that the oceans were filled by comets.¶¶¶¶

Two years after its rebirth, HIFI had answered some of the major questions about the water trail.[15] The water molecules that come out of my tap *are* the same ones that were formed on the surface of a dust grain deep inside a giant molecular cloud. It is also possible again that the Earth's oceans were filled by the comets. The HIFI observations have also shown that there is a lot of water around. It now seems likely that the familiar water molecule, essential to life (at least as we know it), is the second most common molecule in the universe after molecular hydrogen.

We are beginning to see the Lego blocks of life everywhere we look. The Lego blocks for the proteins are the amino acids; the Lego blocks for RNA and DNA are the nucleobases. Both sets have now been found in meteorites and asteroids,[16] and other important bio-molecules have been discovered in molecular clouds.[17] It is beginning to seem quite likely, since its ingredients seem to be everywhere, that life is a common phenomenon. And once it gets started, maybe the evolution of intelligent life is inevitable.

In science fiction, of course, the arrival of an alien spacecraft does not always bode well. But if there are other civilisations out there, it doesn't seem naïve to hope that some of them may already have solved problems we face. Rather than destruction and death, they might arrive bearing solutions – and maybe new forms of art and philosophy, new ways of engaging with the universe.

Of course, we have no way of knowing whether any of this is true. Any respectable scientist is required to state: We just don't know. We now know, which we didn't in the 80s, that virtually every star has planets. There are plenty of places where life *might* exist. But then there is the cosmic lottery problem. If the chance of life arising is incredibly slim, so small that there is only life on one planet in the universe, we must be the winner of the lottery – we are here, after all. Or maybe the universe is brimming with life. Until we find life in at least one other place in the universe - maybe in the clouds of Venus – we have no way of knowing which is true.

But I know which way I'd bet.

¶¶¶¶ Later measurements have shown that not all comets in this class have the same heavy-water proportion as the water in the oceans.[18] It's clearly complicated, but the HIFI observations have shown that the idea that the water in the oceans was delivered by comets is possible again.

CHAPTER 15

The Museum

I N APRIL 2013 ANYONE with observing time scheduled on *Herschel* was biting their nails. The telescope was running on empty. All sub-mm instruments are cooled to ridiculously low temperatures (one of the reasons our branch of astronomy has been so slow to get going), which is the only way to avoid sub-mm radiation from the instrument itself swamping the sub-mm radiation from space. On *Herschel* the main thing keeping the instruments cold was the big tank at the bottom of the cryostat, which contained 335 kilogrammes of liquid helium. Or at least it had at launch. We had always known the helium would not last forever. Every day more of it boiled away, bringing the end of the mission a little closer. The temperature of the detectors in SPIRE was about 0.3 kelvins, less than a degree above absolute zero, but the detectors would only stay this cold while there was helium in the tank. Once all the helium was gone, they would warm up and quickly become useless.

At the beginning of the mission, the predicted lifetime of the helium was 3.5 years, but the uncertainty in that figure was so large we knew it might be six months either way. Towards the end of the mission, someone in the PACS instrument team came up with an ingenious method for checking how much helium was left,* but even with this method it was only possible to predict the end of the mission within a month or so. By April, everyone knew the helium wouldn't last much longer. It might run out this month, it might run out next month – nobody was sure when.

This was a problem for Göran Pilbratt, the *Herschel* Project Scientist. One thing Göran's bosses at ESA headquarters in Paris wanted to avoid was for news of the end of the mission to leak out, which might mean the exhaustion of the helium would be portrayed as some kind of failure by the media. They wanted to mark the end of the mission with a barrage of publicity celebrating its spectacular success, but since nobody knew when this was going

---

* The helium was at a temperature of 1.65 kelvins, but there were additional refrigerators necessary to cool the detectors in PACS and SPIRE down further to their operating temperatures. The instrument teams had to regularly service these refrigerators (obviously remotely). During one of these services, a small amount of heat would be transferred into the big helium tank, which increased the temperature of the helium very slightly. The clever person in the PACS team realised that by measuring this tiny change in temperature, it was possible to estimate how much helium was left in the tank.

DOI: 10.1201/9781003195290-15

to happen there was a problem. The Project Scientist gets all the awkward jobs nobody else knows how to do. The bosses told Göran to sort it all out.

Göran did what all experienced employees of big space organisations do. He wrote a memo. He decided that the crucial moment wasn't when all the helium was gone but when the temperature began to rise. He therefore decided to start the procedure for the end of the mission when the temperature went above 2 kelvins on at least two of the six temperature sensors around the helium tank. In his memo, he stated that when the scientists at the European Space Operations Centre (ESOC) saw this had happened, they should phone him as the *Herschel* Project Scientist. He would then phone ESA headquarters, and the bosses there would make the final decision about the end of the mission. Their decision would then trigger the press releases and other publicity, which had all been arranged in advance.

On 29th April 2013, Operational Day 1446, almost four years after launch, *Herschel* started an observing programme with HIFI. The programme was to observe a sample of dusty supergiant stars, stars that are pumping out chemicals and dust into interstellar space, which I have written about several times in this book. The programme was led by Kay Justtanont of Chalmers University in Sweden. All the observations except one had been successfully completed. The team's very last observation was of the star OH32.8-0.3. A pattern of electrical currents changed and HIFI started the final observation of the programme.

Half an hour later, the observation finished successfully.

There were now a few hours in which there were no scheduled observations, although the helium was still running down because of the need to keep the instruments cold. The end of this period was the start of the daily three-hour contact slot during which the ESA scientists talked to the spacecraft, downloading data and uploading instructions for the next 21 hours. Göran was on a skiing holiday in Austria.

Shortly after the beginning of the contact period, Göran got a phone call from ESOC. It was bad weather and he was in his room at the pensione. The voice on the phone said that the temperatures on the sensors were beginning to rise. It didn't seem likely that the temperature would reach 2 kelvins by the end of the contact period, but it seemed likely that it would exceed the threshold the next day while the telescope was out of contact. Göran decided to call it. He made a phone call to Paris, and within an hour there was an announcement of the end of the mission.

Herschel had made its last observation, but its power was still on. It was still there in space.

The second Lagrangian point, L2, has become a gathering ground for telescopes. *Herschel* went there, so did *Planck*, *Gaia*, the *Wilkinson Microwave Anisotropy Probe*, the *James Webb Space Telescope* and now another ESA mission, *Euclid*. One of its attractions for telescopes like *Herschel* and *Planck* is that at L2, four times further from the Earth than the Moon, the sub-mm radiation from the Earth does not swamp the sub-mm radiation from some faint galaxy or molecular cloud. Its other attraction is that from L2 the Earth and Sun appear in the same place in the sky, which has the advantage that it makes it much easier to avoid pointing the telescope anywhere close to them.[†] The words 'Lagrangian Point' suggest there is a chance all these telescopes might bump into each other, but none is at L2 itself. They are all in orbit around it. *Herschel* itself was on a huge orbit around L2, twice the size of the Moon's orbit around the Earth, which meant that the distance between *Herschel* and the Earth varied between 1.2 and 1.8 million kilometres.

But L2 does have one big disadvantage. An orbit around L2 is not a stable one. Left to itself, a spacecraft put there will start to inexorably drift away from its initial orbit. The ESA engineers had kept *Herschel* on track by occasional small bursts of the spacecraft's rocket engine, but the fuel in that tank, too, wouldn't last forever. Once that fuel was gone, there would be nothing to stop *Herschel* from drifting away.

This probably wouldn't be a problem because the spacecraft would probably drift off into interplanetary space. It probably wouldn't hit the Earth. And if it did, it would probably smack into the ocean or some unpopulated desert. It almost certainly wouldn't hit a city. But there are a lot of 'probablys' in this paragraph….

The ESA engineers were not worried about the spacecraft, which was flimsy enough that they were sure most of it would burn up in the Earth's atmosphere. But they were worried about the telescope's mirror. The mirror was made of silicon carbide, carborundum, one of the toughest substances known. The engineers thought it might survive and hit the ground. It only had a mass of about 300 kilogrammes, but a mass this large hitting the Earth's surface would release as much energy as 3.8 tonnes of TNT. The mirror probably wouldn't hit a city….

Smashing the mirror into Paddington Station or Times Square – take your pick of densely populated places – wouldn't have been great publicity for ESA. Even if the chance of a collision with an inhabited area was infinitesimally small, the engineers couldn't simply turn off the spacecraft's power and forget about it. It was also much too far away to send up another spacecraft to bring it back to Earth. There were only ever two options. One of them was to use the last of the fuel to change the spacecraft's orbit from one around L2 into one around the Sun. But there was another more exciting one: crash it into the Moon.

The first people to suggest this were Göran himself and another ESA scientist, Håkan Svedhem, back in 2010. They had argued that one sure way of avoiding crashing *Herschel* into the Earth would be to crash it into something else. They also argued that it would be a great publicity opportunity (they suggested crashing it into the crater named after William Herschel). During the mission, it became clear there were also some interesting scientific possibilities.

For several decades, almost everything we knew about the Moon was based on the truckload of moon rock, 382 kilogrammes to be precise, hauled back by the Apollo astronauts in the 60s and 70s. Every year now it seems, some spacecraft or other is heading off to the Moon, in preparation for the eventual human return, and these new missions have drastically revised our impression of it. The Apollo rocks contained almost no water, suggesting

---

[†] It's always a bad idea to point a telescope at the Sun because its radiation will fry the instruments. Pointing a sub-mm telescope anywhere close to the Earth is a bad thing because the spillover of sub-mm radiation from the Earth reduces the telescope's ability to detect faint objects.

that the Moon's surface is one of the driest places in the solar system. In 2008, however, an orbiting lunar observatory, the Indian Chandrayaan-1 mission, discovered that in some of the craters close to the Moon's poles, where some of the crater floors are in perpetual shadow, there is ice. NASA's own orbiting lunar observatory, the Lunar Reconnaissance Orbiter, discovered ice in more craters, and the best current estimate is that the ice at the lunar poles contains about 600 billion kilogrammes of water, enough to fill 240,000 Olympic-sized swimming pools.

Water, of course, is one of the main things we will need if we are to return to stay.[‡] The ice in the craters may not be enough for full-scale lunar colonies, but there may also be reserves under the surface. So far, though, we have only been able to learn about the composition of the surface rock. In 2012, as the result of a conversation between Matt Griffin and Nick Bowles, a planetary scientist at Oxford, a team put in a formal proposal to ESA to crash *Herschel* into the Moon. They claimed that by observing the cloud of debris thrown up by the impact with the spectrometers on the Lunar Reconnaissance Orbiter and on telescopes on the Earth, they would be able to find out about the composition of the Moon under the surface.

At the time, the idea raised uneasy feelings in some of us. It possibly didn't feel right, when we are thinking more about the health of our own world, to knock a dent in another.

I also had a problem with the idea of destroying the telescope. I had only ever seen *Herschel* once. When Loretta and I attended the pre-launch meeting at ESTEC in 2007 (Chapter 6), we had the chance to see it before it was shipped out to Korou. We peered through the window of a big cleanroom to see the telescope surrounded by ESA technicians in their blue scene-of-crime suits (Figure 15.1). And now there it was, in interplanetary space. It didn't seem right to smash something that might have historical significance.

It is easy to think, in our online digital world, that objects don't matter. If all the detailed designs of *Herschel* and its instruments are available digitally, why do we need the object itself? But it is not true that this world is eternal. Each time a computer system is updated, the only information transferred from one system to the other is the information somebody values. Even while I have been writing this book, some of the information that was initially available about *Herschel* has vanished. I wrote in Chapter 5 about the clever mechanism designed and built by Pete Hargrave to guide a copper rod through a hole in a box without touching the box – one tiny part of SPIRE. There is a long document describing this mechanism, but if nobody cares to curate it, it will eventually be lost. But if the object itself still exists, this doesn't matter so much.

So if *Herschel* had been crashed into the Moon, I would not have been happy. The ESA managers, however, chose the less exciting option. It was cheaper.[§]

[‡] Give me water and solar power and I can give you a lunar colony. It's easy enough to make oxygen by passing an electric current through the water. An initial reserve of water and oxygen, a good recycling system, power and hydroponics to grow food – these are the basic requirements for a lunar colony.

[§] They claimed that firing *Herschel* into the Moon would cost an extra one million euros. I don't know how they came up with this figure. I am glad about the decision, although I think they should have provided better reasons.

FIGURE 15.1 *Herschel* in the cleanroom at ESTEC.

Credit: ESA

The end of the mission therefore happened behind the scenes. The engineers fired the spacecraft's rocket three times, with the longest burst lasting 11 hours, which emptied the fuel tank and changed its orbit from one around L2 into a stable one around the Sun.

But the spacecraft still had power. Somebody needed to turn it off. In the small group gathered to do this at the European Space Operations Centre, the one who volunteered to press the button was the head of the ESA Science Operations Department, Martin Kessler. The righthand photo in Figure 15.2 shows Martin doing this, in which he is grinning like a Bond villain. But somebody pointed out that this possibly wasn't the right attitude, so they took a second picture, the one on the left, in which Martin is displaying the correct reverence (the one on the ESA website).

Herschel is in a stable orbit around the Sun.[*] There is no air and water out there, so there is nothing to corrode the metal, although every now and then the surface will be hit by a micrometeorite. Nothing else will happen. Unless….

In the Year 2304, the solar system's first quadrillionaire is looking for some way to repair his reputation. The last few years have seen his messy personal life splashed all over the media. He has also made some business decisions, which if he wasn't obviously such an amazingly clever man, might have made him look, frankly, a bit stupid. He doesn't care

---

[*] Apparently, the ESA people who calculate orbits are only certain it will stay in the same orbit for about 300 years, so there is no guarantee that after this it will not hit the Earth. In the context of orbits, stability is a relative thing.

FIGURE 15.2   On the left is the picture of the end of the mission that appeared on the ESA website. On the right is the real picture of the end of the mission.

Credit: ESA and Göran Pilbratt

what people think, of course, but the effect of the media storm on the value of his businesses is in danger of reducing him to trillionaire status. He decides to found a new museum, the last refuge of the disreputable oligarch. He decides the museum will be a museum of space astronomy. It will contain all the space telescopes from the last few hundred years. He dispatches his fleet of spacecraft to bring them all back.

And if *Herschel* finishes its career in a museum on the Moon being gawked at by children on school trips, that wouldn't be a bad ending.**

I don't remember hearing about the end of the mission myself. I imagine I heard about it casually in the corridor or at coffee. I don't suppose I reacted much. We had known from the beginning that *Herschel* would only have a limited life and, besides, a lot of us were still working on our data from the telescope.

All the observations ever made by *Herschel* are now in its archive. I have always found it surprising how many of the observations in telescope archives have never been touched, but astronomers are often involved in so many observing programmes that it is hard to reduce the data from every observation. On the positive side, the deluge of data from a modern telescope like *Herschel* means that its archive is a glittering treasure trove to search through.††

I and the rest of the *Herschel* ATLAS team, though, did still care about our data. We were still trying to transform the raw data we had received from the telescope into sub-mm images and catalogues, which we still planned to release to the community. It was just taking such a long time.

** Apparently, the possibility of eventually bringing the spacecraft back was discussed in ESA.
†† Archive trawling is a lot of fun and easier than writing a proposal.

By the time the helium ran out, Loretta and Steve had been in New Zealand for a year and a half. After the end of the mission, for about 18 months we carried on as we had before. Elisabetta, Matt and I worked on the data in Cardiff, with regular Skype calls to Loretta and Steve in New Zealand. Every summer we had a face-to-face meeting on their annual visit to the UK. Progress was very slow.

As they had expected, Loretta and Steve loved the scenery of New Zealand, but they found it much harder to do research in astronomy there. They also missed their families and friends.

Sometime in 2014, they decided to make a change. A friend of ours, Rob Ivison, another Brit who was then Director of Science at the European Southern Observatory, had been awarded a large grant by the European Research Council which allowed him to hire several postdocs. Rob, also a member of the *Herschel* ATLAS team, offered Loretta and Steve one of the postdocs between them. They made a deal with their university that they would spend six months each year in New Zealand teaching their courses and then six months back in Europe paid by Rob's research grant. To complete this picaresque story, Rob was working at the headquarters of the European Southern Observatory, which is in Munich, but he still had a house in Edinburgh. During their six months in the northern hemisphere, Loretta and Steve worked in Edinburgh and lived in the basement of Rob's house. Rob is a very flexible fellow.

It was all a little complicated, but at least they were back in the UK for six months every year and it made working together much easier. This went on for a bit until Rob's grant ran out of money. By this time, Loretta and Steve had decided to move back to the UK permanently. Another Cardiff professor and member of the *Herschel* ATLAS team, Haley Gomez, also had a research grant from the European Research Council. She offered both Steve and Loretta postdocs, and they came back to Cardiff.

This made things much easier. Towards the end of 2016, which was now seven years since we had seen the first images, it began to seem we might soon release the images and catalogues for our three small fields. We found ourselves again combing through catalogues of tens of thousands of galaxies looking for inconsistencies. We were back staring at the screens again, arguing over which blob was part of some big galaxy and which was a separate source. We began to inch towards the edge of the cliff.

Just before Christmas 2016 we finally did it – nobody wanted this to drag into the new year – and we released to the worldwide astronomy community, years later than I had promised in our proposal, the images and catalogues for the three small fields.

It turns out that once you've jumped off a cliff, doing it again is quite easy. Once we had released the first set of images and catalogues, a lot of the anxiety vanished. Everyone was happy we had finally released them, and nobody immediately spotted any flaws in what we had done. We did a second data release for our other fields in 2017 and a third one in 2022.

And the survey became fun again. I finally did something I should have done almost a decade earlier.

I have mentioned several times now that the galaxies discovered by a survey in the visible waveband, the traditional kind of survey that has been done since the time of the Herschels, seem to fall in two general classes. I explained in Chapter 11 that a good way to show this is to plot the galaxies in a diagram of colour versus luminosity (how much radiation the galaxy emits rather than how much we detect on the Earth). In that chapter I showed a cartoon version of the diagram (Figure 11.1). Figure 15.3 shows this diagram for a real survey.[1] There were so many galaxies found by this survey, roughly 21,000, that if I plotted each galaxy as a point in the diagram, all the points would merge into a single splurge of black. Instead, I have divided the diagram into squares, using the darkness of each square to represent the number of galaxies found in the survey that fall in this square (the darker the square, the more galaxies in the square). Most of the galaxies clearly fall in two main areas. The streak at the top is the red sequence, the rounder area below is the blue cloud and the area between, where there are fewer galaxies, is the green valley.

Within a couple of months of our first *Herschel* observations, Loretta and Steve and their team in Nottingham had managed to identify over two thousand of the sub-mm-emitting galaxies (Chapter 10). So it would have taken me only a few minutes to see where the galaxies discovered in a sub-mm survey fall in the same diagram. Eight years later, I finally got around to doing it.

Figure 15.4 shows the result. This time, I have used contours to represent the number of galaxies from the survey in the visible waveband that fall in different parts of the diagram (higher contours, more galaxies). The two contour mountains show, as in the previous diagram, that the galaxies fall in two areas: the red sequence and the blue cloud, with the green valley, the gap between the two mountains, in between. In this figure I have used the darkness of each square to show the number of galaxies discovered in our sub-mm survey that fall in that square. The galaxies discovered in the two surveys clearly don't lie in the same places. The galaxies discovered in the two surveys clearly don't lie in the red sequence or the blue cloud. They have in-between colours, in an area I decided to call, in

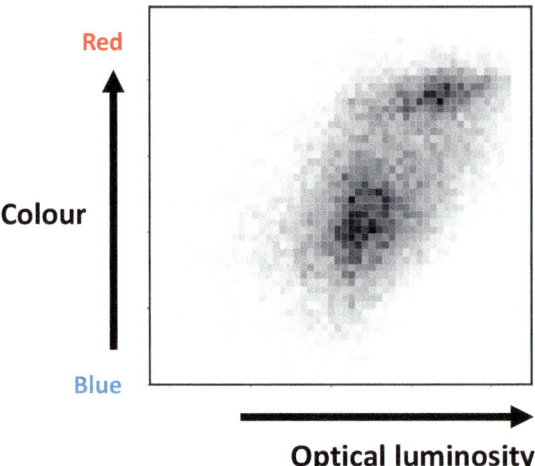

FIGURE 15.3 Colour versus optical luminosity for 21,000 galaxies discovered in a survey in the visible waveband.[1] The diagram is divided into small squares, with the darkness of each square representing the number of galaxies that fall in that square.

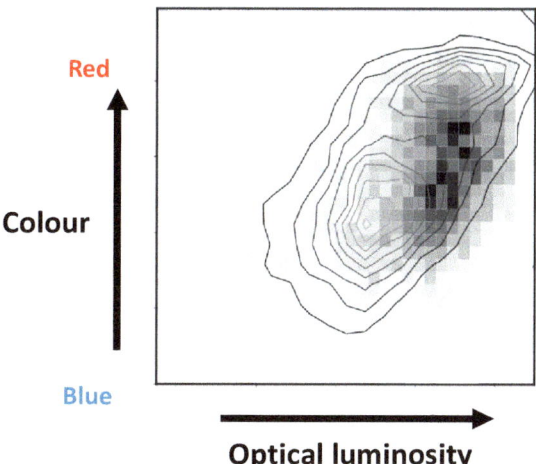

FIGURE 15.4 The same plot of colour versus optical luminosity as in the previous figure, with the areas occupied by the galaxies discovered in the survey in the visible waveband now shown by contours. The darkness of each small square in this figure represents the number of galaxies discovered in our sub-mm survey that fall in that square.

homage to all the other names, the 'green mountain'.[2] When Loretta, looking at the pictures of our galaxies on her laptop in Sydney Airport, thought they didn't look the same as the galaxies discovered in one of the traditional surveys, she had been right.

For 20 years we had been wondering whether the galaxies discovered in a sub-mm survey would be different from those discovered in a survey in the visible waveband. Now we had the answer. The universe through sub-mm spectacles *does* look different to the universe seen through optical spectacles.

To find out what this all meant, I had to turn to another *Herschel* survey.

While we had all been working on the *Herschel* ATLAS, I had also been part of another team that was carrying out a survey using some of the observing time awarded to the instrument builders, the guaranteed time (Chapter 6). In designing this other survey, the *Herschel* Reference Survey, we tried to do something that never happens with an astronomy survey: create a fair, unbiased sample of galaxies.

Any survey in a single waveband is necessarily biased. A survey in the visible waveband does not produce a fair, unbiased sample of galaxies because it preferentially finds ones that are luminous in the visible waveband, which in practice means galaxies that contain many of the short-lived, massive stars that dominate the visible light from most galaxies. A sub-mm survey is not any better because it preferentially finds galaxies that are luminous in the sub-mm waveband, which in practice means ones that contain a lot of dust. A survey in any waveband is biased towards the galaxies that are luminous in that waveband.

Another problem with a survey in a single waveband is the range of distance of by the galaxies in the survey. More luminous galaxies can be detected further away, which means there is often a huge variation in the distances. The galaxies discovered in the *Herschel*

ATLAS ranged from ones that are almost next door to ones that are so far away we are seeing them less than one billion years after the big bang.[3]

With the *Herschel* Reference Survey, we tried to do something better. In the last 200 years, so many surveys have been made, in so many different wavebands, that most of the massive galaxies in the nearby universe are already in one catalogue or another. We decided to choose a volume of space, use the existing catalogues to compile a list of all the massive galaxies in that volume, and then take pictures of all of them with *Herschel*. There were 323 massive‡‡ galaxies in the volume of space we chose. Of these, 65 were elliptical and lenticular galaxies. The rest were spirals and irregulars. Ellipticals and lenticulars don't usually contain much dust, so we didn't expect the sub-mm pictures of them would be very exciting. But that was the point of the project. We wanted to take pictures of all the galaxies in this volume, not just the pretty ones, a sample that was as close as we could get to being a representative sample of galaxies.[4]

I think it was me who came up with the name of this survey. I called it 'Herschel' after the telescope and 'Reference' because I envisaged it being a reference book for astronomers in the future. But I was excited to learn later, when I checked while I was writing this book, that the name was better than I had imagined because 196 of the galaxies in the Herschel Reference Survey were discovered by the Herschels themselves.[6]

I've shown some of their drawings of the galaxies side-by-side with our sub-mm images of the same objects (Figure 15.5). In this book, I have tried to make an imaginative connection between astronomy as it is done today and as it was done in the time of the Herschels, but these pictures show the gulf between the two times. When we made our observations, we uploaded a file of positions onto a website, which someone later uploaded to the telescope during one of the contact periods. The spacecraft then took pictures of the galaxies in robot mode – I often didn't even know when one of our observations was being made. William drew his pictures from what he could see through the telescope with hands that must often have been stiff with the cold.

Drawing a picture would also not have been as simple as it sounds. William was seeing nebulae at the limits of human perception and the Herschels' observing strategy was designed to keep his eyes adapted to the dark. William stayed up on the observing platform in the dark, while Caroline, in a nearby room, wrote down what he saw and calculated the positions of the nebulae (Chapter 7). To draw a picture of a nebula, William would have had to light a candle, which would have ruined his ability to see the nebula.

They seem to have overcome this problem in the following way.§§ At the foot of the telescope was a workman, who, by cranking on a windlass, was moving the telescope slowly up or down. When William discovered a nebula he wanted to sketch, he called down to the

---

‡‡ I am simplifying a bit, because we actually used a different mass limit for ellipticals and spirals (Chapter 6).[5]

§§ I have obtained much of this (with a little supposition from myself) from *William Herschel: Discoverer of the Deep Sky* by Wolfgang Steinicke, which is a must-read for anyone interested in the Herschels' observing methods.

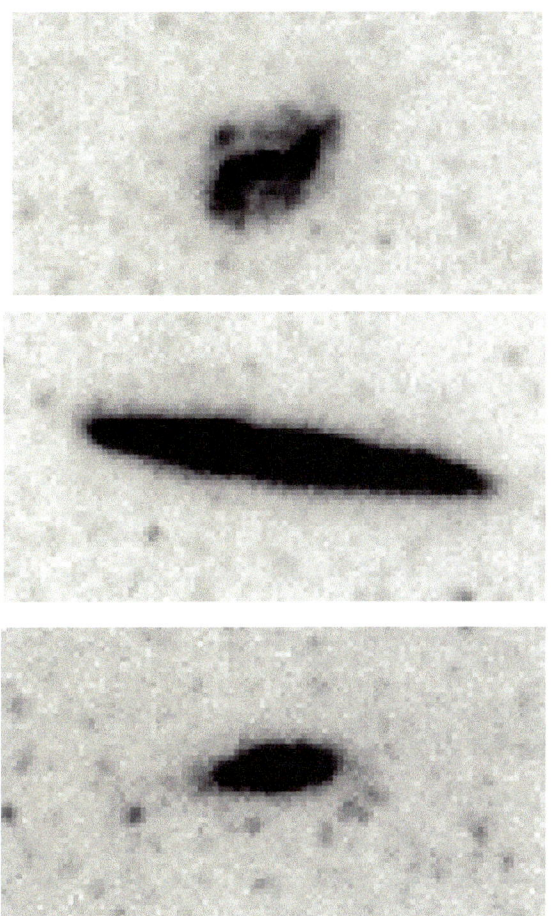 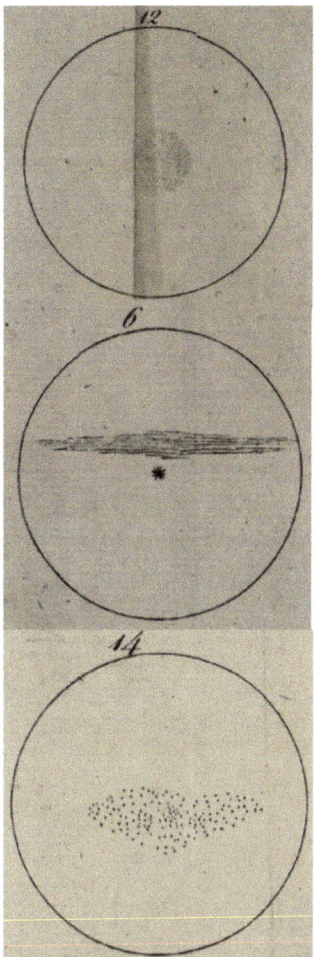

FIGURE 15.5   In the left-hand column are three of our sub-mm images. In the right-hand column are drawings of the same galaxies sketched by William Herschel in the 1780s. The faint dark blobs in our images are the sub-mm galaxies seen in every sensitive *Herschel* image.[7]

Credit: ESA and the *Herschel* Reference Survey team

workman to stop the cranking. Up on the platform, William would still have been able to move the telescope a small amount in an east-west direction. As the nebula drifted through the field-of-view, William, by moving the telescope, would have been able to keep it in view for about three minutes before it drifted completely out of sight.[8] During this time, I suspect he looked at it through several interchangeable eyepieces, since ones with different magnifications would have brought out different aspects of the nebula's structure. Once he thought he had studied it enough, he lit a candle and made his sketch. Since the candle's light probably immediately destroyed the dark adaptation of his eyes, making the nebula temporarily invisible, I suspect he must have done these sketches from memory.

Given these challenges and that the pictures are in different wavebands, it is surprising that William's drawing look as similar to our images as they do.⁵⁵

One of the advantages of such a well-studied sample of galaxies is that it is possible to estimate some of a galaxy's most important properties. Sometime after we finished the *Herschel* Reference Survey, Pieter de Vis, a postdoc who was working with Loretta and Steve at the University of Canterbury in New Zealand, estimated the stellar birth rate and the mass of all the stars in the galaxy for all 323 galaxies.[9] I had recently agreed to supervise some undergraduate students for a research project. When I heard about Pieter's work, I suggested to the students that they plot a diagram that showed the normalised stellar birth rate*** of each of the galaxies in the *Herschel* Reference Survey versus the total mass of all the galaxy's stars. When they came back to show me the diagram a couple of weeks later, I thought they had made a mistake.

What I had expected to see was a diagram like the one in Figure 15.6, which illustrates what this diagram usually looks like for the galaxies discovered in a survey in the visible waveband. As in the diagram of colour versus luminosity (Figure 15.3), the galaxies in this new diagram mostly fall in two areas. The strip along the top contains galaxies with a high stellar birth rate, the same galaxies that are in the blue cloud in the previous diagram. The circular area at the bottom contains the galaxies that were in the red sequence in the previous diagram. The correspondence between the diagrams is not surprising because the colour of a galaxy is a good indication of the number of short-lived massive stars it

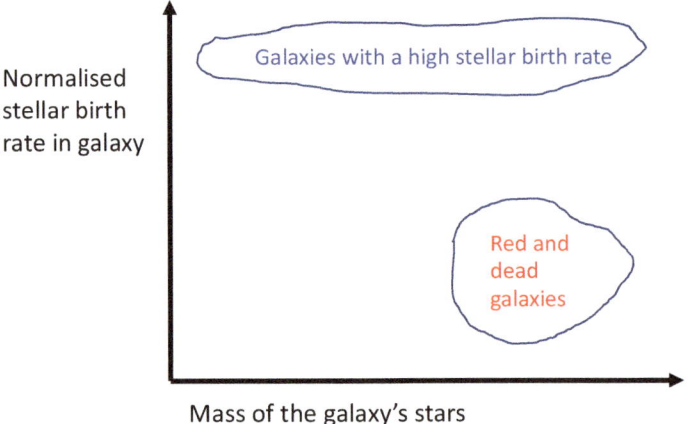

FIGURE 15.6   A typical plot of normalised stellar birth rate in a galaxy versus the total mass of the galaxy's stars for the galaxies discovered in a survey in the visible waveband.

---

⁵⁵   William's sketches show the galaxies' starlight. The sub-mm pictures show the galaxies' dust. The similarity of the pictures (obtained 200 years apart) shows that the distributions of the dust and stars in these galaxies are quite similar.

***   The normalised stellar birth rate is the stellar birth rate divided by the mass of all the galaxy's stars, which is a way of making a correction for the stellar birth rate being high simply because the galaxy is big (as the number of births in a hospital may be high simply because it's a big hospital).

contains, which is a good measure of its stellar birth rate – blue means a high stellar birth rate, red means a low one.[†††]

As in the previous diagram, the important feature is the gap between the two areas (the green valley in the previous diagram). As I explained in Chapter 13, the natural evolutionary process for a galaxy is for it to move, as its reservoir of gas is gradually used up and its stellar birth rate falls, from the area of blue galaxies at the top of the diagram to the area of red galaxies at the bottom. But this natural process is quite slow, and if this were all there was, we should see far more galaxies in the in-between area than we do. As I explained in that chapter, astronomers have come up with a lot of ideas for some process that might move a galaxy quickly enough across this gap so that few galaxies are seen there. The most popular idea today is that it is the monster lurking at the centre of a galaxy. It seems plausible, although the details of how this happens are not understood, that as gas starts to fall into the central supermassive black hole, the energy produced is enough to transform a galaxy from one with a large gas reservoir and a high stellar birth rate into a galaxy with a depleted gas reservoir and a low stellar birth rate - and rapidly enough that we see few galaxies during the transformation.

When the students showed me the diagram they had plotted for the galaxies from the *Herschel* Reference Survey, I thought they had made a mistake. But when I checked, they had done everything correctly. In their diagram (Figure 15.7), there was no gap.[10]

After thinking about it for a bit, I realised the explanation. The *Herschel* Reference Survey was a fair sample of the universe (at least as far as we could make it) but any survey in a single waveband is always biased. When I looked at the properties of the galaxies from the *Herschel* Reference Survey in Figure 15.7 that filled in the gap seen in Figure 15.6, I

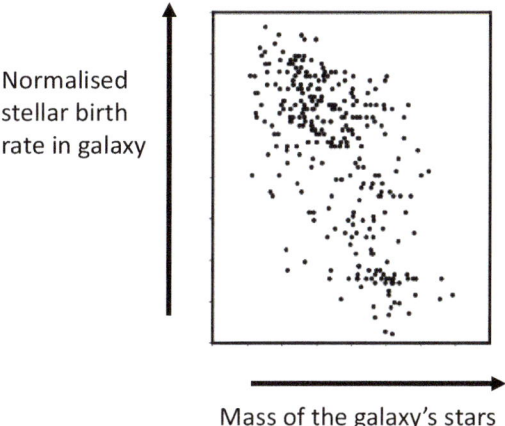

Normalised stellar birth rate in galaxy

Mass of the galaxy's stars

FIGURE 15.7   Normalised stellar birth rate in a galaxy versus the total mass of all the galaxy's stars for the galaxies in the *Herschel* Reference Survey.[10]

[†††] The shapes of the areas are not the same – the red sequence in one diagram turns into a circular area in the other – because the relationships between colour and luminosity in one diagram and normalised stellar birth rate and total stellar mass in the other are not simple ones.

found that they contained an unusually large amount of dust. Any survey in the visible waveband would therefore tend to miss galaxies like this because the dust hides the visible light, which explains the gaps in both Figures 15.3 and 15.6. Any sub-mm survey, on the other hand, would preferentially find galaxies like this, which explains the green mountain in Figure 15.4.[11] I realised that the blue cloud, the red sequence and the green valley (and the green mountain) are the result of the biases inherent in surveys carried out in a single waveband.

The discovery that the gaps in these diagrams are the result of survey bias has an important consequence. Astronomers have written hundreds of papers about processes that might speed galaxies across the gaps in these diagrams. If there is no gap, there is no need for any of these. The evolution of galaxies may not need the monster at the centre of a galaxy to come to life. It might be a much gentler process than anyone thought. I don't expect everyone to accept this overnight, but that is what the data seems to say.

Whether our analysis is correct or not, it does show the danger of relying on observations in a single waveband. Visualism is wrong. As in society, so in astronomy – alternative points of view are important.

A space mission never really ends.

Astronomers around the world are now using the images and catalogues on which Loretta and I and the rest of the *Herschel*-ATLAS team spent a decade of our lives. There is no sub-mm telescope in space and there is unlikely to be one for another decade, so what we created is likely to be useful for at least that long. And probably longer – astronomers are still using data from the IRAS mission in the 80s.

A large part of a space mission is its people. Some of the people I wrote about in this book have moved on. Elisabetta Valiante moved with her partner to Vancouver, where she now works for a quantum computing company. Mattia Negrello became a lecturer in Cardiff but has now returned to Italy for a non-astronomy adventure. Loretta and Steve decided they didn't want to return to the pressure of permanent academic jobs. They decided to stay postdocs and do research for as long as possible. Eventually the research grants ran out, and they are now practising sustainable living in the Welsh countryside (last week I saw on social media that Loretta now has chickens). Göran Pilbratt, the *Herschel* Project Scientist, was moved, in the ESA way, onto other projects. While still working on the aftermath of *Herschel*, he briefly worked on the planet-finder mission *PLATO* and then on another mission to observe other planetary systems, *Ariel*. He is now formally retired but he is still called back by ESA to review space missions. Everyone else I mentioned in the book is still active in astronomy. Matt Griffin has the office on the floor above me and is still planning new space telescopes.

As for me, I spent two decades using telescopes on the ground before I joined the *Herschel* team. *Herschel* had rockets, launches, orbits, space jargon ('nominal', 'good to go') – all the excitement of space without going there myself. Since the helium ran out, I have been on the lookout for another space mission to join. I am now part of another

ESA mission: *Euclid*. The pictures taken by this new space telescope will show almost the same detail as *Hubble's* but will cover almost one-third of the sky. *Euclid* was launched on a Falcon 9 rocket from Cape Canaveral on 1st July 2023. We are now taking data.

A space mission never really ends. *Herschel* is still there orbiting the Sun. It will be there forever unless somebody goes out and brings it back. The helium may be gone and the power off, but we still have its archive, the memory of everything it saw. The telescope itself would be a good centrepiece for a museum, but the rest of the museum is already complete, every drawer and cabinet packed with observations of fantastical beasts.

The End.

Acknowledgements

I WOULD LOVE TO THANK everyone I worked with during the *Herschel* mission, who made it such a great adventure, but there are too many and I don't want to leave anyone out. However, I must single out Matt Griffin, who got me involved in *Herschel* and whose warmth and humanity were two of the reasons the *Herschel* mission was so much fun. I didn't work so closely with the *Herschel* Project Scientist, Göran Pilbratt, during the mission, but his personaility and can-do attitude were also reasons why the mission was so enjoyable. I must also thank Loretta Dunne for her part in our 20-year quest to do the first big submillimetre survey of the sky, and the students and postdocs in Cardiff who worked with me on the *Herschel* ATLAS and the *Herschel* Reference Survey: Robbie Auld, Ali Dariush, Simon Dye, Michael Pohlen, Matt Smith and Elisabetta Valiante. I am also grateful to the many colleagues (again too many to name individually) in the School of Physics and Astronomy at Cardiff University who have provided a supportive environment for my research.

I would also like to thank all of those who agreed to be interviewed by me after the end of the mission: Maarten Baes, Mike Barlow, Dave Clements, Gianfranco de Zotti, Loretta Dunne, Simon Dye, Andy Fabian, Walter Gear, Reinhard Genzel, Jane Greaves, Peter Hargrave, Matt Griffin, Mikako Matsuura, Mattia Negrello, Stephane Ott, Nicolas Peretto, Göran Pilbratt, Michael Rowan-Robinson, Stephen Serjeant, Matt Smith and Elisabetta Valiante.

I also received a lot of help from other professional astronomers, who despite the competition in our field are generally a very supportive lot. Several people in the astronomers' Facebook group, in particular Chris Conselice and Ray Norris, made helpful suggestions about the history of our understanding of interstellar dust, which substantially improved Chapter 3. Wolfgang Steinicke is a historian of science rather than an astronomer, but his comments substantially improved Chapter 7. I am grateful to Maarten Baes, Mike Barlow, Walter Gear, Jane Greaves, Mikako Matsuura, Carole Tucker and Erwine van Dishoeck for comments on individual chapters. I am exceptionally grateful to Simon Dye, Matt Griffin and Göran Pilbratt for reading and commenting on the whole thing (Matt Griffin twice). Any errors that remain in the book are, of course, my own. Finally, a big thank you to Matt Smith for constantly helping me out with software, figures and finding my phone.

At CRC, I am grateful to Haylie Allan, Carolina Antunes and Betsy Byers. I am especially grateful to Betsy for her comments on all the chapters (many times twice), which were both useful for improving the text and for encouraging the author.

This book was started when I was a Visiting Fellow at Clare Hall in Cambridge. I am grateful to the college but also to a group of visiting academics who made Keirsten's and my time in Cambridge so much fun: Kevin Edwards, Amy Livingstone, John Murphy, Lynn Pepall, Andrew Ramage, Nancy Ramage, Dan Richards, Li Tang and Evan Zimroth.

I am grateful to my children, Juliet, Nicholas and Oliver for their supportive comments (and jokes). Finally, overwhelmingly, I am grateful to my wife Keirsten. She read the penultimate draft of the entire book, and her comments led to so many improvements that I wished I had given it to her earlier.

Notes and References

CHAPTER 1

The historical background to Horizon 2000 and *Herschel* comes from books and the interviews. The two books that I found most useful were:

A History of the European Space Agency, 1958–1987, Volumes 1 and 2 (ESA Publications Division)
Inventing a Space Mission – The Story of the Herschel Space Observatory (Springer)

The last of these, a formal academic history of the mission, provides all the technical details of the observatory and contains a detailed description of the tribulations of the mission in the 90s, which would otherwise have been lost to history. The timeline for the last few months before launch comes from blogs that were written at the time by the ESA scientists. I have reconstructed launch day itself from my own memories and those of the others that were there. If the reader is surprised by my perfect recall, after a ten-year gap, of every detail of what was happening to the spacecraft in the minutes after launch, the answer is: YouTube. The story of all the emotions of that day, though, is based on all our memories.

1   A phrase attributed to James Carville, the main political strategist in Bill Clinton's successful 1992 presidential campaign, although according to Wikipedia he never actually said this.
2   *Inventing a Space Mission*, 31
3   *Inventing a Space Mission*, 58–59

CHAPTER 2

The story of the observing run on Mauna Kea is a reconstruction from memory and the details in reference [6] below.

1   Herschel, W. 1800, *Philosophical Transactions of the Royal Society of London*, 90, 284
2   Noon, K. and De Napoli, K. 2022, *First Knowledges – Sky Country*, Thames and Hudson
3   Hamacher, D. et al. 2023, *Journal of Archaeological Science*, 159, 105819
4   Bennett, A.S. 1962, *Memoirs of the Royal Astronomical Society*, 68, 163
5   Hewish, A. et al. 1968, *Nature*, 217, 709
6   Eales, S.A. et al. 1989, *Astrophysical Journal*, 339, 859

CHAPTER 3

1   I have taken most of the biographical information about Barnard and Ranyard from *The Immortal Fire Within* (IFW) by William Sheehan
2   IFW, 3
3   IFW, 211
4   IFW, 77, 172, 212
5   IFW, 317
6   IFW, 4

7 Steinicke, W. 2016, *Journal of Astronomical History and Heritage*, 19, 305–326

8 Norris, R.P. and Hamacher, D.W. 2013, *Handbook of Archaeoastronomy and Ethnoastronomy*, volume 2, edited by C.L.N. Ruggles, Springer Reference

9 IFW, 151

10 Barnard, E.E. 1890, *Monthly Notices of the Royal Astronomical Society*, 50, 310

11 *Knowledge*, November 1st 1889

12 Barnard, E.E. 1916, *Astrophysical Journal*, 43, 1

13 IFW, 172

14 IFW, 71

15 Russell, H.N. 1922, *Nature*, 110, 81

16 Trumpler, R.J. 1930, *Publications of the Astronomical Society of the Pacific*, 24, 214

17 Ménard, B. et al. 2010, *Monthly Notices of the Royal Astronomical Society*, 405, 1025; Smith, M. et al. 2016, *Monthly Notices of the Royal Astronomical Society*, 462, 331

18 Brownlee, D. 2014, *Annual Reviews of Earth and Planetary Science*, 42, 179

19 Westphal, A.J. et al. 2014, *Science*, 345, 786; Brownlee, D. 2014, *Annual Reviews of Earth and Planetary Science*, 42, 179

20 Heck, P.R. et al. 2020, *Proceedings of the National Academy of Sciences of the United States of America*, 117, 1884

21 Dwek, E. and Krennrich, F. 2013, *Astroparticle Physics*, 43, 112

22 *Colours of the Galaxies* by David Malin and Paul Murdin, Promotional Reprint Company

CHAPTER 4

1 Eales, S. et al. 1999, *Astrophysical Journal*, 515, 518; Lilly, S. et al. 1999, *Astrophysical Journal*, 518, L641.

2 Hughes, D.H. et al. 1998, *Nature*, 394, 241.

3 Chapman, S. et al. 2005, *Astrophysical Journal*, 662, 772

4 Smail, I. et al. 1997, *Astrophysical Journal*, 490, L5

5 Not the same galaxies. The descendants of the sub-mm galaxies are galaxies that today are out of sight (because of the speed of light), but we think that they must be the same kind of elliptical galaxies we do see around us today.

6 Dunne, L. et al. 2000, *Monthly Notices of the Royal Astronomical Society*, 315, 115; Dunne, L. and Eales, S. 2001, *Monthly Notices of the Royal Astronomical Society*, 327, 697

CHAPTER 5

Most of the history in this chapter comes from *Inventing a Space Mission* and from my interviews with Walter Gear, Reinhard Genzel, Matt Griffin, Pete Hargrave, Stephan Ott, Göran Pilbratt and Michael Rowan-Robinson. To keep the footnotes as few as possible, I have only given them where it is not obvious how I got the information.

1 Wilson, R.W. et al. 1970, *The Astrophysical Journal*, 161, L43

2 The list is taken from Table 2 of Tielens, A. 2013, *Reviews of Modern Physics*, 85, 1021

3 Pilbratt, G. et al. 2020, *The Herschel Space Observatory Development, Operation and Post-operations: Lessons Learned, Proceedings of the SPIE*, 11443, 09

4 Interview with Reinhard Genzel

5 Interview with Reinhard Genzel

6 Interview with Reinhard Genzel

7 Interview with Matt Griffin

8 Interviews with Reinhard Genzel, Göran Pilbratt and Michael Rowan-Robinson

9 Interview with Reinhard Genzel

CHAPTER 6

1 Conselice, C. et al. 2016, *Astrophysical Journal*, 830, 83
2 Tremonti, C.A. et al. 2004, *Astrophysical Journal*, 613, 898
3 Simon, J.D. et al. 2016, *Astrophysical Journal*, 838, 11

CHAPTER 7

The biographical matter for this chapter is mostly from two wonderful books: *The Age of Wonder* (AW) by Richard Holmes and *Discoverers of the Universe – William and Caroline Herschel* (DU) by Michael Hoskin. Most of the information about the techniques of the Herschels' survey – fascinating for an astronomer – comes from another wonderful book: *William Herschel – Discoverer of the Deep Sky* (WH) by Wolfgang Steinicke.

All the extra detail is from William Herschel's own papers. The more imaginative sections of the chapter are mostly based on these papers, and in these sections I have often used William's own words. Since this book is not an academic history, I haven't bothered with footnoting everything, and I usually only give a note where there is something debatable or where the source is not one of the books above. The notes at the ends of the more fanciful paragraphs list the sources on which it is based.

William Herschel's papers must be read with an awareness that science then was very different. In one paper, for example, he makes confident assertions about the existence of people on the Moon and the Sun which seem to my eyes as a modern astronomer to be based on wild leaps of reasoning. But it must be remembered that to make any scientific progress at all, given the limited available observational and experimental apparatus at the time, it was essential to be quite daring with the arguments and conjectures made from the data. Herschel himself makes this point (reference b below):

> If we indulge a fanciful imagination and build worlds of our own, we must not wonder at our going wide from the path of truth and nature….On the other hand, if we add observation to observation, without attempting to draw not only certain conclusions, but also conjectural views from them, we offend against the very end for which only observations ought to be made. *I will endeavour to keep a proper medium; but if I should deviate from that, I could wish not to fall into the latter error* (my italics).

A nice illustration of this is a comparison of Herschel's papers on nebulae - galaxies as we know them now - and on the inhabitants of the Sun. Both contain the same methods of reasoning, the same daring conjectures and arguments, and the same bold assertions. But the papers on galaxies now seem amazingly prescient, the paper on the Sun-people completely bonkers.

The papers of Herschel that I have used in this chapter are as follows:

(a) Account of Some Observations Tending to Investigate the Construction of the Heavens, 1784, Herschel, W., *Philosophical Transactions of the Royal Society of London*, 74, 437
(b) On the Construction of the Heavens, 1785, Herschel, W., *Philosophical Transactions of the Royal Society of London*, 75, 213
(c) Catalogue of One Thousand New Nebulae and Clusters of Stars, 1786, Herschel, W., *Philosophical Transactions of the Royal Society of London*, 76, 457
(d) Catalogue of a Second Thousand of New Nebulae and Clusters of Stars; With a Few Introductory Remarks on the Construction of the Heavens, 1789, Herschel, W., *Philosophical Transactions of the Royal Society of London*, 79, 212
(e) On Nebulous Stars, Properly So Called, 1791, Herschel, W., *Philosophical Transactions of the Royal Society of London*, 81, 71
(f) On the Nature and Construction of the Sun and Fixed Stars, 1795, Herschel, W., *Philosophical Transactions of the Royal Society of London*, 85, 46

(g) On the Power of Penetrating into Space by Telescopes; With a Comparative Determination of the Extent of That Power in Natural Vision, and in Telescopes of Various Sizes and Constructions, 1800, Herschel, W., *Philosophical Transactions of the Royal Society of London*, 90, 49

(h) Catalogue of 500 New Nebulae, Nebulous Stars, Planetary Nebulae, and Clusters of Stars; With Remarks on the Construction of the Heavens, 1802, Herschel, W. *Philosophical Transactions of the Royal Society of London*, 92, 477

1 This paragraph and the next are based mostly on Herschel's own words -AW, DU, f

2 The Quaker is named in AW as John Michell, who was the first to propose the existence of black holes. Although John Michell did own a reflecting telescope that William bought after his death (DU), he was also an Anglican clergyman and so could not have been the Quaker from whom William bought his original tools.

3 AW, DU, f – Reference f, written many years later, gives an account of his arguments for why the Moon, the Sun and the other stars, planets and moons must contain life, although he acknowledged that it must be very different from life on Earth. This paper also contains a good example of the style of analogical reasoning he used in all his scientific work.

4 AW, DU

5 DU

6 AW

7 DU, c

8 c

9 c

10 DU

11 WH

12 a

13 DU, a, b

14 The method the Herschels used for mapping the Galaxy is described in b. In this paper, William gives the distance to the Galaxy's boundary in units of the distance to Sirius, which he thought, mistakenly, to be the closest star because it's brightest. It is only in a much later paper (h) that he uses an estimate for the distance of Sirius (about six light-years) to give an estimate of the distance to the boundary in physical units. Although there were some mistaken assumptions in the method, the Herschels' estimate of the extent of the Galaxy is similar to modern estimates of the thickness of the Galaxy's disk, which is what they would have measured given that the diagram in Figure 7.2 is actually a cross-section of the Galaxy that is roughly perpendicular to the disk (Figure 7.3).

15 WH

16 DU

17 a

18 The account of this night is based on reference a. I needed to give myself a little creative license because in the paper William does not give the location of this rich nebulous bed. However, it seems likely that it is the Virgo cluster, the nearest rich cluster of galaxies, which in spring is overhead at midnight. Nebula hunting would have been best in spring because that is when the local supercluster, the association of galaxy clusters of which Virgo is part, is in the night sky. Among his many other discoveries, William was the first to notice that galaxies often come in clusters.

19 AW

20 a, b

21 This paragraph is based on a paper written many years later (h), which contains an estimate of the distance out to which the nebulae could be seen with the forty-foot telescope. This estimate is based on the assumption that a typical nebula contains 50,000 stars. On this assumption, William estimated that he could see nebulae out to about two million light-years. I have scaled this estimate by the ratio of the diameters of the mirrors of the two telescopes to get an estimate of the distance out to which the Herschels would have been able to see nebulae with the smaller telescope they used for their survey, although both estimates are much too low because real galaxies contain many more stars than 50,000.

22 d

23 a, b, d – '*If we indulge a fanciful imagination and build worlds of our own, we must not wonder at our going wide from the path of truth and nature... On the other hand, if we add observation to observation, without attempting to draw not only certain conclusions, but also conjectural views from them, we offend against the very end for which only observations ought to be made.*' (ref b)

24 Oliver Wendell Holmes quoted in DU

25 e

26 f

27 DU, f

28 Goldman, W. *Adventures in the Screen Trade*

CHAPTER 8

1 Although a planetary nebula looks like a planet, Herschel correctly realised that it is radiation from a diffuse fluid in interstellar space. He thought that the nebula might eventually contract to form a star and thus be connected to the birth of a star rather than its death: 'When we reflect upon these circumstances, we may conceive that, perhaps in progress of time these nebulae which are already in such a state of compression, may be still farther condensed so as actually to become stars.' Herschel, W., 1811, *Philosophical Transactions of the Royal Society of London*, 101, 269

2 Wesson, R. et al. 2010, *Astronomy and Astrophysics*, 518, L44

CHAPTER 9

1 'When we reflect upon these circumstances, we may conceive that, perhaps in progress of time these nebulae which are already in such a state of compression, may be still farther condensed so as actually to become stars.' Herschel, W., 1811, *Philosophical Transactions of the Royal Society of London*, 101, 269

2 Eales, S. et al. 2012, *Astrophysical Journal*, 761, 168

3 Men'shchikov, A. et al. 2010, Astronomy and Astrophysics, 518, L103

4 Andre, Ph. et al. 2010, *Astronomy and Astrophysics*, 518, L102

5 Arzoumanian, D. et al. 2011, *Astronomy and Astrophysics*, 529, L6

6 Estimate from Conselice, C. et al. 2016, *Astrophysical Journal*, 830, 83

CHAPTER 10

1 Smith, M. et al. 2017, *Astrophysical Journal Supplement*, 233, 26

2 Valiante, E. et al. 2016, *Monthly Notices of the Royal Astronomical Society*, 462, 3146

3 There was some evidence for low-redshift evolution from much earlier studies but nothing so convincing (e.g. Lonsdale, C. et al. 1990, *Astrophysical Journal*, 350, L6; Eales, S. 1993, *Astrophysical Journal*, 404, 51; Lawrence, A. et al. 1996, *Monthly Notices of the Royal Astronomical Society*, 308, 897). Our results are in Dye, S. et al. 2010, *Astronomy and Astrophysics*, 518, L10

CHAPTER 11

1 Walsh, D. et al. 1979, *Nature*, 279, 381
2 Negrello, M. et al. 2007, *Monthly Notices of the* Royal *Astronomical Society*, 377, 1557
3 Negrello, M. et al. 2010, *Science*, 330, 800
4 The ALMA Partnership 2015, *Astrophysical Journal*, 808, L4
5 Swinbank, A. et al. 2015, *Astrophysical Journal Letters*, 806, L17
6 Dye, S. et al. 2015, *Monthly Notices of the Royal Astronomical Society*, 452, 2258
7 Thomas, D. et al. 2010, *Monthly Notices of the Royal Astronomical Society*, 404, 1775

CHAPTER 12

1 Hafez, I. 2010, *Abd al-Rahman al-Sufi and His Book of the Fixed Stars – A Journey of Re-discovery*, PhD thesis, James Cook University.
2 Herschel, W. 1785, *Philosophical Transactions of the Royal Society of London Series I*, 75, 213
3 Hubble, E.P. 1929, *Astrophysical Journal*, 69, 103
4 Attempts to map out the spiral structure of Andromeda are described in two papers: Gordon, K.D. et al. 2006, *Astrophysical Journal*, 638, L87; Kirk, J.M. et al. 2015, *Astrophysical Journal*, 798, 58. I am still not convinced.
5 Block, D.L. et al. 2006, *Nature*, 443, 832
6 Lewis, A. et al. 2015, *Astrophysical Journal*, 805, 183
7 Smith, M.W.L. et al. 2012, *Astrophysical Journal*, 756, 40
8 Planck Collaboration 2014, *Astronomy and Astrophysics*, 571, A11; Tabatabaei, F.S. et al. 2014, *Astronomy and Astrophysics*, 561, A95; Hunt, L.K. et al. 2015, *Astronomy and Astrophysics*, 576, A33
9 Ford, G.H. et al. 2013, *Astrophysical Journal*, 769, 55
10 Kirk, J.M. et al. 2015, *Astrophysical Journal*, 798, 58

CHAPTER 13

Most of the account of the discovery of Supernova 1987A is based on a long interview that Ian Shelton gave when he returned to Chile from Canada in 1987, which is on YouTube: https://www.youtube.com/watch?v=1GLAvXqSzos.

1 Dye, S. et al. 2010, *Astronomy and Astrophysics*, 518, L10
2 The Illustris Project: https://www.illustrisproject.org/static/illustris/media/illustris_hubble_diagram.jpg
3 Eales, S.A. et al. 2018, *Monthly Notices of the Royal Astronomical Society*, 473, 3507
4 Dunne, L. et al. 2011, *Monthly Notices of the Royal Astronomical Society*, 417, 151
5 Eales, S.A. et al. 2012, *Astrophysical Journal*, 761, 168
6 Wesson, R. et al. 2015, *Monthly Notices of the Royal Astronomical Society*, 446, 2089
7 Matsuura, M. et al. 2011, *Science*, 333, 1258

CHAPTER 14

1 Sagan, C. 1961, *Science*, 133, 849
2 Morowitz, H. 1967, *Nature*, 215, 1259
3 Greaves, J. et al. 2021, *Nature Astronomy*, 5, 655
4 Greaves, J. et al. 1998, *The Astrophysical Journal*, 506, L133
5 Smith, B.A. and Terrile, R.J. 1984, *Science*, 226, 1421
6 Acke, B. et al. 2012, *Astronomy and Astrophysics*, 540, 125
7 de Vries, B.L. et al. 2012, *Nature*, 490, 74
8 Halliday, T. *Otherlands – A World in the Making*, Penguin Books
9 van Dishoeck, E. et al. 2014, *Protostars and Planets VI, University of Arizon Press*, 835
10 Caselli, P. et al. 2012, *The Astrophysical Journal Letters*, 759, L37
11 Tielens, A. 2013, *Reviews of Modern Physics*, 85, 1021
12 Hogerheijde, M. et al. 2011, *Science*, 334, 338
13 Figure 5 in van Dishoeck, E. et al. 2014, *Protostars and Planets VI*, University of Arizon Press, p835
14 Hartogh, P. et al. 2011, *Nature*, 478, 218
15 van Dishoeck, E. et al. 2021, *Astronomy & Astrophysics*, 648, A24
16 Cooper, G.W. and Cronin, J.R. 1995, *Geochimica et Cosmochimica Acta*, 59, 1003; Oba, Y. et al. 2023, *Nature Communications*, 14, 1292; Paschek, K. et al. 2023, *Astrophysical Journal*, 942, 50
17 Gorai, P. et al. 2021, *Astrophysical Journal*, 907, 108; Mifsud, D.V. et al. 2021, *Frontiers of Astronomy and Space Sciences*, 8, 213
18 Altwegg, K. et al. 2015, *Science*, 347, 387

CHAPTER 15

1 The sample of galaxies came from the Galaxy and Mass Assembly project. Details of the sample are given in Eales, S. et al. 2018, *Monthly Notices of the Royal Astronomical Society*, 481, 1183
2 Eales, S. et al. 2018, *Monthly Notices of the Royal Astronomical Society*, 481, 1183
3 Fudamoto, Y. et al. 2017, *Monthly Notices of the Royal Astronomical Society*, 472, 2028
4 Boselli, A. et al. 2010, *Publications of the Astronomical Society of the Pacific*, 122, 261
5 Eales, S.A. et al. 2017, *Monthly Notices of the Royal Astronomical Society*, 465, 3125
6 I discovered which of the galaxies in our sample had been discovered by the Herschels by consulting the tables on the website of Dr. Wolfgang Steinicke: www.klima-luft.de/steinicke/index_e.htm.
7 The images on the left are *Herschel* images at 250 micrometres of, from top to bottom, the galaxies NGC 4123, NGC 4517 and NGC 4592. The sketches of the same galaxies on the right were published in Herschel, W. 1784, *Philosophical Transactions of the Royal Society of London*, 74, 437, with the names of the galaxies, which were not given in the paper, shown in Figure 2-45 of *William Herschel: Discoverer of the Deep Sky* by Wolfgang Steinicke.
8 *William Herschel: Discoverer of the Deep Sky* by Wolfgang Steinicke.
9 De Vis, P. et al. 2017, *Monthly Notices of the Royal Astronomical Society*, 464, 4680
10 Eales, S. et al. 2017, *Monthly Notices of the Royal Astronomical Society*, 465, 3125
11 Eales, S. et al. 2018, *Monthly Notices of the Royal Astronomical Society*, 481, 1183

Index